ANATOMY

of

STRETCHING

General Disclaimer

The contents of this book are intended to provide useful information to the general public. All materials, including texts, graphics, and images, are for informational purposes only and are not a substitute for medical diagnosis, advice, or treatment for specific medical conditions. All readers should seek expert medical care and consult their own physicians before commencing any exercise program or for any general or specific health issues. The author and publishers do not recommend or endorse specific treatments, procedures, advice, or other information found in this book and specifically disclaim all responsibility for any and all liability, loss, or risk, personal or otherwise, which is incurred as a consequence, directly or indirectly, of the use or application of any of the material in this publication.

Thunder Bay Press
An imprint of Printers Row Publishing Group
10350 Barnes Canyon Road, Suite 100, San Diego, CA 92121
www.thunderbaybooks.com

Printers Row Publishing Group is a division of Readerlink Distribution Services, LLC.
Thunder Bay Press is a registered trademark of Readerlink Distribution Services, LLC.

All notations of errors or omissions should be addressed to Thunder Bay Press, Editorial Department, at the above address. All other correspondence (author inquiries, permissions) concerning the content of this book should be addressed to Moseley Road, Inc., 123 Main Street, Irvington, NY 10533. www.moseleyroad.com.

Thunder Bay Press
Publisher: Peter Norton
Publishing Team: Lori Asbury, Ana Parker, Kathryn Chipinka, Aaron Guzman
Editorial Team: JoAnn Padgett, Melinda Allman, Dan Mansfield
Production Team: Jonathan Lopes, Rusty von Dyl

ISBN: 978-1-68412-090-1

Printed in China

21 20 19 18 17 1 2 3 4 5

ANATOMY
of
STRETCHING

A Guide to Increasing Your Flexibility

Craig Ramsay

with a foreword by
Jerry Mitchell

THUNDER BAY
P·R·E·S·S

San Diego, California

CONTENTS

Foreword by Jerry Mitchell	6
Introduction: Stretching for a Better Life	8
Build a Stretching Routine	10
Full-Body Anatomy	18
The Stretching Session	**20**
Assisted Foot Stretches	22
Point	22
Flexion	22
Slope-Down	23
Slope-Up	23
Band-Assisted Stretches	24
Wing Stretch	24
Sickle Stretch	25
Seated Leg Cradle	26
Unilateral Seated Forward Bend	27
Bilateral Seated Forward Bend	28
Butterfly Stretches	30
Seated Butterfly	30
Folded Butterfly	31
Scoop Rhomboids	32
Front Deltoid Towel Stretch	33
Lying-Down Arch Stretch	34
Lying-Down Groin Stretch	35
Lying-Down Pretzel Stretch	36
Unilateral Leg Stretches	38
Unilateral Knee-to-Chest Stretch	38
Unilateral Leg Raise	39
Hip Adductor Stretch	40
Lying-Down Figure 4	42

Internal Hip Rotator Stretch	44
Happy Baby Stretch	46
Side-Lying Rib Stretch	48
Side-Lying Knee Bend	50
Cobra Stretch	52
Back Stretches	54
Child's Pose	54
Kneeling Lat Stretch	55
Cat Stretch	55
Pigeon Stretch	56
Shin Stretch	58
Frog Straddle	60
Half Straddle Stretches	62
Half Straddle	62
Side-Leaning Half Straddle	63
Double-Leg Straddle Split	64
Chest-to-Thigh Straddle Split	66
Chest-to-Floor Straddle Split	68
Toe Touch	70
Standing Back Roll	71
Good Morning Stretch	72
Scalp and Facial Stretches	74
Scalp Stretch	74
Lion Stretch	75
Eye Box Stretch	75
Neck Stretches	76
Side Neck Tilt	76
Downward Neck Tilt	76
Upward Neck Tilt	77
Neck and Head Turn	77
Back-of-the-Neck Stretch	77
Triceps Stretch	78
Biceps Stretch	79
Wall-Assisted Chest Stretch	80
Forearm Stretches	82
Wrist Flexion	82
Wrist Extension	83

Calf Stretches	84
Calf Heel Drop	*84*
Toe-Up Calf Stretch	*85*
Standing Quadriceps Stretch	86
Kneeling Sprinter Stretch	87
Sumo Squat	88
Side-Leaning Sumo Squat	90
Side-Lunge Stretch	92
Forward Lunge	94
Forward Lunge with Twist	96
Straight-Leg Lunge	98
Downward-Facing Dog	100
Wide-Legged Forward Bend	102

Partner Stretches | **104**
Assisted Butterfly Stretch	106
Assisted Happy Baby	107
Assisted Unilateral Thigh Stretch	108
Assisted Unilateral Leg Raise	110
Assisted Chest Stretch	112
Assisted Seated Forward Bend	114
Assisted Child's Pose	116
Assisted Pretzel Stretch	117
Russian Split Switch	118

Pregnancy Stretches | **120**
| Torso Rotation | 122 |
| Hand-on-Knee Stretch | 124 |

Lying Pelvic Tilt	125
Unilateral Good Morning Stretch	126
Cat Stretch	127
Downward-Facing Dog	128

Office Stretches | **130**
Workplace Stretch Routine	131
Seated Twists	132
Seated Figure 4	133
Forward Bend Hip Shift	134
Double-Leg Hinge	135
Supported Hamstrings Stretch	136

Foam Roller Stretches | **138**
Tennis Ball Foot Massage	139
ITB Roll	140
Foam Roller Lat Stretch	142
Foam Roller Back Stretch	143
Calf and Hamstrings Stretch	144
Foam Roller Shin Stretch	146

Extreme Challenge | **148**
Dancer's Lunge	150
Lying-Down Side Hamstring Stretch	151
Bilateral Quad Stretch	152
Front Split	153
Russian Splits	154
Russian Split	*154*
Roll-Through from Russian Split	*155*
Standing Extensions	156
Assisted Side Tilt	*156*
Standing Leg Extension	*157*

The Quick Stretch Program | **158**

| Credits and Acknowledgments | 160 |
| About the Author | 160 |

FOREWORD

Just saying the word *stretch* makes me want it to last forever. Stre-e-e-e-e-e-etch! Elongate! Expand! Lengthen!

To me, it's a word that means "make it last." Make it last for as long as possible.

Have you ever watched dogs when they awaken from a wonderful night of rest or a midday nap? They rise on all fours and before a minute passes, they're in a downward-dog position or some other fantastic move. They're stretching their bodies, waking up their muscles, getting ready for full-out movement.

Stretching is the first thing I do in the morning and last thing I do before I fall asleep. For me, stretching is one of the most important things I do on a daily basis. And I do a lot! I wish I did more. My body feels like I am giving it a wake-up call to snap to attention and prepare for the inconceivable things I will ask it to do today.

As a Broadway dancer, choreographer, and director, I am always in the rehearsal room inventing musicals. The shows I like to work on require, for the most part, lots of movement. Stretching before every session helps prepare the body for what lies ahead. A quick bend, jump, twirl, split, or kick without stretching first would mean a quick trip to the emergency room for me.

So what makes stretching so fantastic? I'll leave that expertise to Craig Ramsay.

I first became aware of Craig and his amazing body when he was appearing in the Broadway revival of *Fiddler on the Roof*. Watching a large, very muscular guy move with so much grace and flexibility, needless to say, caught my eye. I spend countless hours, days, weeks, and years working with some of the most amazing dancers, who all stay in top-notch physical condition. That's Craig! He's extremely flexible and that's directly connected to his muscle growth, maintenance, longevity, and stretching program.

Anatomy of Stretching will answer all your stretching questions. Craig will share all the technical reasons why you should stretch. Emotionally, what I can tell you is this: No one should go through a day without some form of stretching. It will make you feel better. It will make your body feel better. It's just that simple.

No one has to tell a dog to stretch. It is in their DNA to do it— and to do it before anything else. Stretching will lengthen not only your muscles, but also your mind and spirit. And in turn, your life. Yes, I believe stretching your muscles can lengthen your life. And it can, with the proper diet and exercise, make you look a whole lot better.

So what are you waiting for? Make it last! Let's stre-e-e-e-etch!

Jerry Mitchell

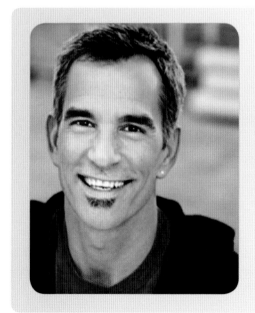

JERRY MITCHELL

Jerry Mitchell's choreography has been seen on Broadway in *Catch Me if You Can, Legally Blonde* (which he also directed), *Dirty Rotten Scoundrels, The Full Monty, Hairspray, La Cage* (TONY Award), *Never Gonna Dance, Gypsy, The Rocky Horror Show,* and *You're a Good Man, Charlie Brown.* For London's West End, he also choreographed *Legally Blonde* (which he also directed), *Hairspray,* and *Love Never Dies.* Jerry is also the conceiver/producer of Broadway Bares, an annual charity event for Broadway Cares, and producer/director/choreographer of its long-running Las Vegas counterpart, Peepshow.

STRETCHING FOR A BETTER LIFE

As a child, I had difficulty staying still. When doctors diagnosed my problems as attention-deficit/hyperactivity disorder (ADHD), they tried to convince my parents that the *only* way to keep me focused, still, and attentive would be through medication.

Medication has been proven to be effective with certain ADHD cases, but thanks to my parents, I'm a firm believer in utilizing sports and other physical activity to combat the inattention and hyperactive or impulsive behavior associated with ADHD.

My parents took notice that I could stay still and calm when I was trying to impress them with a gymnastics move or an incredibly demanding stretch. So, during family events, my dad would often encourage me to put on a show of my talents. Oh, the hours I needed to perfect my back walkover and my impressive splits. My dad knew that all that practicing was quiet time for his overenergized son. My mom also realized that stretching kept me focused, so she started to encourage me to do my homework sessions on the living room floor while doing my hour of routine stretching.

Soon my parents enrolled me in dance, gymnastics, and hockey—anything and everything that would keep me physically active and out of trouble. Their plan worked, too—my marks in school improved, as did my self-confidence.

I have a great deal of love for this book and its message, because I truly believe the time I put into my stretching helped me to manage my ADHD and boosted my physical capabilities, which enabled me to become such an asset on the ice rink, on school teams, and, most important, on the stage. It's no exaggeration to say that stretching helped me to achieve my successful Broadway dancing, singing, and acting career. It also helped to turn me into one of the most sought-after fitness experts in the world.

My hope is that the stretches and routines that follow go far beyond helping you stay agile and in shape. I hope that you are open to this book helping you to overcome whatever obstacles life may throw at you.

Craig Ramsay

BUILD A STRETCHING ROUTINE

While traveling on airplanes, we hear the safety advice to put on our own oxygen masks before helping loved ones or neighbors. In this case, taking care of yourself first is a necessity.

The same rule applies to your stretching session or exercise routine—and to life itself for that matter. Do not feel guilty for taking care of yourself. Make achieving self-worth and a healthy body new goals. You are no good to your loved ones if you are in poor health.

Types of Stretching

There are many forms and types of stretching, such as static, dynamic, passive, and active. You should be aware of the options, but segregating the types of stretching is not the most beneficial way of executing a stretching routine. Combine all of forms of stretching for an ideal routine.

Static stretching consists of stretching a muscle to its farthest point and then holding that position.

Dynamic stretching consists of controlled movements to increase a particular body part's range of movement.

Isometric stretching, a type of static stretching, calls for the resistance of muscle groups through isometric contractions (tensing) of the stretched muscles.

Active stretching consists of assuming a position and then holding it with no assistance other than using the strength of your agonist, or opposing, muscles.

Passive stretching consists of an external force (either a person or an apparatus) bringing the joint of a relaxed person through its full range of motion.

Proprioceptive neuromuscular facilitation (PNF) stretching, originally developed as a form of physical therapy, combines passive and isometric stretching. It usually calls for a 10-second push phase followed by a 10-second relaxation phase.

This book calls for mostly static, passive, and active stretches. These forms of stretching are extremely important for anyone looking to gain flexibility and develop body awareness.

A Full-Body Stretch

There are so many opinions on what constitutes the ideal order of stretches and how stretching sessions should be designed, but limited data exists to promote any one method. The Stretching Session included in this book (pages 20–103) is the result of taking into consideration a great deal of information from both the field of dance and the field of personal training. It covers all of the major muscle groups and their opposing muscles, offering a well-rounded stretch routine that guides you to discover your strengths and weaknesses. What are your problem areas? Which are your tightest muscles? Is one side stronger than the other? Just how flexible are you? What is holding you back from your full living and sports potential?

The Stretching Session begins at the feet—and for good reason. The ankles

ADVICE FOR THE ATHLETE

A muscle must be flexible enough to have a slightly greater range of motion than what your sport requires, but not so much more that it diminishes your performance by becoming too loose and out of control.

and feet are the base of your body. With 26 bones, the foot is a complex structure, and, as a dancer does, you should address this area first. The health of your feet can affect your whole body, including internal organs. Tight feet and ankles lead to tight hamstrings, calves, and hips. Addressing foot flexibility issues leads to stronger cardiovascular capabilities and lowers the risk of leg strain and pain during exercise.

Many of the stretches in this book target the back, the inner thighs, and the hamstrings, all areas that are susceptible to tightness. The hamstrings are a group of three muscles located at the back of the leg.

These muscles—the biceps femoris, semitendinosus, and semimembranosus—are the major knee flexors. The hamstrings are typically the tightest muscles in the lower body.

One or Both Legs?
The nervous system's protective blockage begins much sooner bilaterally (both legs at the same time) than unilaterally (one leg at a time). The range of motion is therefore more restricted. We must not forget to stretch unilaterally, as this form of stretching increases range of motion. Ask any dancer and she will admit that we all have one leg that is more flexible for high kicking than the other.

How Often Should You Stretch?
In a perfect world, you could stretch every day; unlike weight training, you don't need a rest day in between sessions to heal and repair the muscles. Aim to make stretching part of your daily routine. Don't, however, beat yourself up when you don't fulfill your daily stretching goals. In this imperfect world, every day just might not work for you. Just include extra "me time" in the next day's schedule. Keep the commitment to yourself, just as you would be sure to keep a commitment to a client, or as you would never miss your child's sporting event.

WORKING WARM

Your body works best when your internal temperature is high: your muscles are warmer and therefore more relaxed, your quick response time is up, your perceived exertion is low, and your heart rate is lower.

Find out when your body temperature is typically at its highest. Most of us peak in the late afternoon, with our morning temperatures the lowest of the day; therefore, stretching in the afternoon may be ideal.

Important note: If the weather is very cold, or if you are feeling very stiff, take extra care to warm up properly before you stretch in order to reduce your risk of injury.

Between sets, light stretching of your target muscles enhances a strength-training workout. If you are working on your biceps, try the Biceps Stretch (page 79).

Performing a stretching session four or more times a week will make an incredible impact on your life. Even doing it once a week will yield benefits in ways that will surprise you.

Once you start stretching regularly, you'll feel those benefits. Remember that even when you feel stressed, going through your stretches is a good thing. I've worked with clients with stressful lives who sometimes go through the stretching session featured in this book twice a day. It helps, and it works.

No Excuses!

Don't waste time finding excuses for why you can't stretch. Learn the stretches in this book so that you can perform them anywhere and everywhere. You will find the appropriate stretches to perform:

- first thing in the morning, and last thing at night;
- in front of the television during family movie night or during your favorite one-hour TV program;
- with a friend;
- during long periods in front of the computer;
- while taking a short break between loads of laundry;
- during a long plane flight in the comfort of your seat;
- and anytime you are feeling stiff, sore, or stressed!

Just remember to fulfill your new goal of taking care of yourself, and fit in the stretches where and when you can.

BENEFITS OF STRETCHING

The benefits of stretching are many. Here are a few:

Increases flexibility and energy. Becoming more flexible can improve your physical performance and decrease your risk of injury. A flexible muscle is far more resistant to injury than an inflexible one. Stretching produces a slight rise in muscle-tissue temperature; this raises the point at which the fiber breaks. Stretching improves the efficiency of energy-generating enzymes that can provide you with more energy during your workout.

Burns calories. An extensive dynamic/active stretching program can help you increase your rate of calorie burn.

Improves cardiorespiratory endurance. Stretching helps develop the body's ability to supply fuel and oxygen during sustained physical activity. It also raises your fatigue threshold.

Combats the effects of aging. As you get older, you lose flexibility, but a proper stretching program can help you regain and maintain it. Improving circulation to your muscles shortens recovery time if you have suffered muscle injuries. Stretching helps you achieve your healthiest range of joint motion, which keeps you in better balance. Better balance is important as you grow older, because it decreases the chance of injury from falls.

Relieves stress. Stretching relaxes the tense, tight muscles that are often associated with stress. You can use a stretching regimen to alleviate emotional disruptions and improve your focus and concentration.

Improves muscle coordination. Regular stretching reduces the time it takes for messages to travel from the muscles to the brain.

Relieves lower-back pain. Stretching your hip flexor, hamstrings, and gluteal muscles, along with the lumbar region of your spine, increases the range of movement in your pelvis and lumbar spine. This increased mobility reduces lower-back pain.

Elongates muscles. Longer muscles have greater growth potential. Combining a stretching program with a proper weight-resistance workout will help you develop larger, more impressive-looking muscles.

Taking time for stretching means taking time for yourself. Taking time for yourself is liberating—it can help you clear your mind, gain self-confidence, and organize and focus on your goals and aspirations. It can also help you sleep better, and a well-rested you is better equipped to achieve your goals!

When Not to Stretch: Lessons from a Dancer

When building a stretching routine, it is crucial to find the appropriate balance between muscle tension and muscle flexibility. All cardiovascular and weight-resistance exercises need a proper stretch and contraction. Think of a muscle as an elastic band: overstretch it and it loses its ability to properly contract, thus decreasing its strength.

A fitting example of improper stretching technique comes from an unlikely source: dancers. Unfortunately, some dancers misuse their extraordinary flexibility when preparing for a cardiovascular or strength-training workout. Quite a few dancers warm up with excessive stretching prior to a workout.

Yet, however impressive the ability to perform a full straddle split may be, the key to a successful stretching routine is knowing when to go for the big stretches and when to focus on the smaller ones. Overstretched muscles respond like loose noodles, leaving them susceptible to injury. Dancers with overstretched muscles limit the weight they can lift, and this limitation impairs their form.

Learn from these dancers' mistakes. Light stretching is acceptable before and during

WHEN TO STRETCH

A stretching program can alleviate or prevent the following muscle conditions or injuries.

Muscle cramping. Muscle cramping results from poor hydration, a diet low in magnesium and sodium, incorrect positioning, and/or inappropriate movement or exercise. Excessive alcohol intake, diabetes-related problems, and narrowing of the arteries due to plaque buildup are also known to cause cramping. Nighttime cramps have been linked to deficiencies in B vitamins, magnesium, and calcium. To relieve or eliminate a cramp, stretch the cramped muscle to force it to immediately relax.

Stiff muscles. A day or two after intense exercise, you may experience stiff muscles, which can last from a few days to more than a week. Stretching lightly during workouts, after workouts, and during the days following a workout can prevent or lessen muscle stiffness.

Muscle spasms. A muscle spasm is a symptom of fatigue and may signal possible injury. It usually shows up as a painful knot in the muscle. Spasms can take days to control and resolve. To lessen and even prevent muscle spasms, stretch the area in which you most often experience spasms prior to and during exercise.

Pulled muscles. Stretching a muscle too forcefully or beyond its normal length can result in an extremely painful injury known as a pulled muscle. Rest is absolutely necessary to heal a pulled muscle, but stretching properly is the best way to prevent one.

Muscle tears. A muscle tear can occur if you push yourself too hard when your muscles are already injured or extremely fatigued. A torn muscle is a serious injury—it can sideline an athlete for up to six months, for example, and even worse, it can affect an athlete's entire career.

Muscle ruptures. A muscle rupture occurs when a large group of fibers inside the muscle is damaged. You are at the highest risk for a rupture when your muscles are fatigued, and you then push them beyond their limits. Combining a stretching program with a proper bodybuilding or weight-resistance program is the best way to protect your muscles from possible ruptures.

exercise; for example, during strength training, lightly stretching your target muscles between sets loosens the muscles and makes for a better pump. But remember to save your intense stretching routine for *after* your workout.

A Proper Warm-up

To get the most from your stretching session, you need to warm up properly. And keep in mind that stretching is not warming up! For example, every Broadway show has a "half-hour" call, warning all the performers that they have 30 minutes to get into their places. But responsible Broadway dancers know they need more than a half hour to prepare, physically and mentally. They know they need to leave ample time for a head-to-toe warm-up before they even begin the stretches that will keep them limber for a performance.

You should begin as dancers do, rotating all of your joints to get them lubricated, starting with your toes and working up to your fingers.

1 Wiggle all of your toes for 5 to 10 seconds.
2 Rotate each ankle 5 to 10 times.
3 Bend each knee for 5 to 10 seconds.
4 Rotate your hips for 5 to 10 rotations on each side.
5 Twist your torso back and forth for 5 to 10 seconds.
6 Rotate each shoulder 5 to 10 times.
7 Bend each elbow for 5 to 10 seconds.
8 Rotate your neck, and move it from side to side for 5 to 10 seconds.
9 Rotate each wrist for 5 to 10 seconds.
10 Wiggle all of your fingers for 5 to 10 seconds.

Make slow, clockwise movements, and then reverse to counterclockwise for each joint.

After the joint rotations, you should perform about 5 to 8 minutes of aerobic/cardiovascular activity, which increases the oxygen available to the body and enables your heart to use oxygen more efficiently. Increasing blood flow in the muscles not only improves muscle performance and flexibility but also reduces the chance of injury.

Low-tech aerobic warm-up exercises include:

- Bouncing or running in place
- Jogging
- Jumping rope
- Bouncing on a small trampoline
- Toe taps on a Bosu ball
- Jumping jacks

High-tech cardio warm-up machines include:

- Treadmills
- Rowing machines
- Stationary bicycles
- Stair climbers
- Elliptical machines

Start your aerobic warm-up at 40 percent of your maximum heart rate, which should feel like an easy pace, and then progress to about 60 percent of your maximum heart rate.

LEARN FROM THE DANCERS

Nearly every dancer has one leg that is more flexible than the other. This "good leg" gets used repeatedly, such as when kicking impressively at auditions or on the Broadway stage. Yet this reliance on the good leg creates an imbalance that affects posture and can even lead to injury. Muscles become tight and overdeveloped on one side—and can end a dancer's career early. Other people can experience the same kind of imbalance with daily repetitive movements. This book can help you identify where those issues may exist—knowledge you can then address with your health-care professional.

Warming up benefits stretching by improving your coordination, increasing elasticity, and raising your level of body awareness.

Cooling Down

You can use the Quick Stretch Program (pages 158–59) at the end of your workout as a cooldown to the Stretching Session.

A proper cooldown in your workout program offers a number of benefits, including:

- Lowers your heart rate and breathing to a normal rate.
- Keeps your range of motion and flexibility intact.
- Helps avoid dizziness or fainting, which can sometimes result from the sudden cessation of vigorous activity.
- Reduces the immediate postexercise tendency for muscle spasms, cramping, and stiffness (especially in women).
- Maintains the blood flowing throughout the body.
- Helps reduce muscle injury, stiffness, and soreness.

What to Wear

Have you heard the saying, "If you dress the part, you will feel the part"? Make it a goal to look your best when you engage in an exercise program.

- Wear clothes that flatter your physique and boost your confidence. If you work out in a gym or health club, make sure that they are suitable for the setting. Choose breathable fabrics that move well and don't constrict you.

- Layer your clothing. Keeping warm is very important during all physical activity, especially stretching. Taking off a layer of clothing once your body is warmed up takes a lot less time than healing a muscle that has been stretched cold.

- Wear shoes appropriate for your workout, and replace them when necessary. For instance, purchase a new pair of running shoes after 300 miles—or at the first noticeable sign of wear and tear.

- Take care of your personal hygiene: brush your teeth, shower regularly, and wear deodorant.

- Avoid wearing excessive makeup during a workout.

- Keep your hair simple and neat.

If you look your best, you'll feel your best, and if you feel your best you'll put more into your workout. Put your all into your workouts, and in a surprisingly short time you will achieve incredible results, reaching your health and physique goals.

Your Daily-Living Workout

Daily life can place strenuous demands on our bodies. To prevent injury, stretch regularly to improve your posture. A stretching routine teaches you to move more efficiently and with purpose, making your daily chores and habits, such as carrying groceries, getting in and out of the car, holding and rocking the baby to sleep, climbing stairs, or picking up clothes to do a load of laundry, far more efficient. All of these activities, when performed with good posture, will burn more calories, giving you, in effect, a "daily-living workout."

The Body-Mind Connection

This book will help you build an awareness of the muscles engaged during a variety of stretches, which will also help you to better control the way your body moves through space. You'll get to know your body and start moving it with intention, efficiency, and purpose. Stretching slows you down enough to put you on the path toward becoming one in mind, body, and spirit.

FULL-BODY ANATOMY

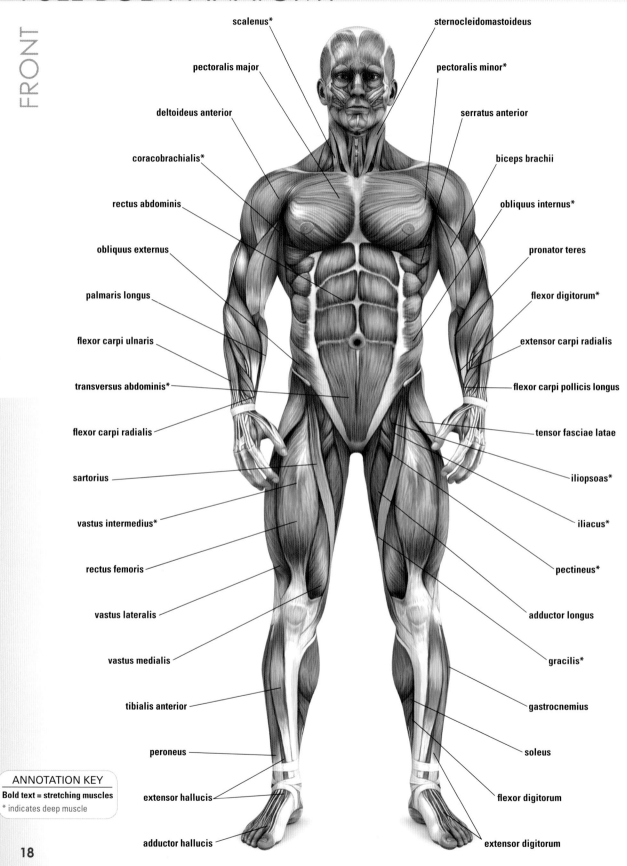

scalenus*

sternocleidomastoideus

pectoralis major

pectoralis minor*

deltoideus anterior

serratus anterior

coracobrachialis*

biceps brachii

rectus abdominis

obliquus internus*

obliquus externus

pronator teres

palmaris longus

flexor digitorum*

flexor carpi ulnaris

extensor carpi radialis

transversus abdominis*

flexor carpi pollicis longus

flexor carpi radialis

tensor fasciae latae

sartorius

iliopsoas*

vastus intermedius*

iliacus*

rectus femoris

pectineus*

vastus lateralis

adductor longus

vastus medialis

gracilis*

tibialis anterior

gastrocnemius

peroneus

soleus

extensor hallucis

flexor digitorum

adductor hallucis

extensor digitorum

> ANNOTATION KEY
> **Bold text = stretching muscles**
> * indicates deep muscle

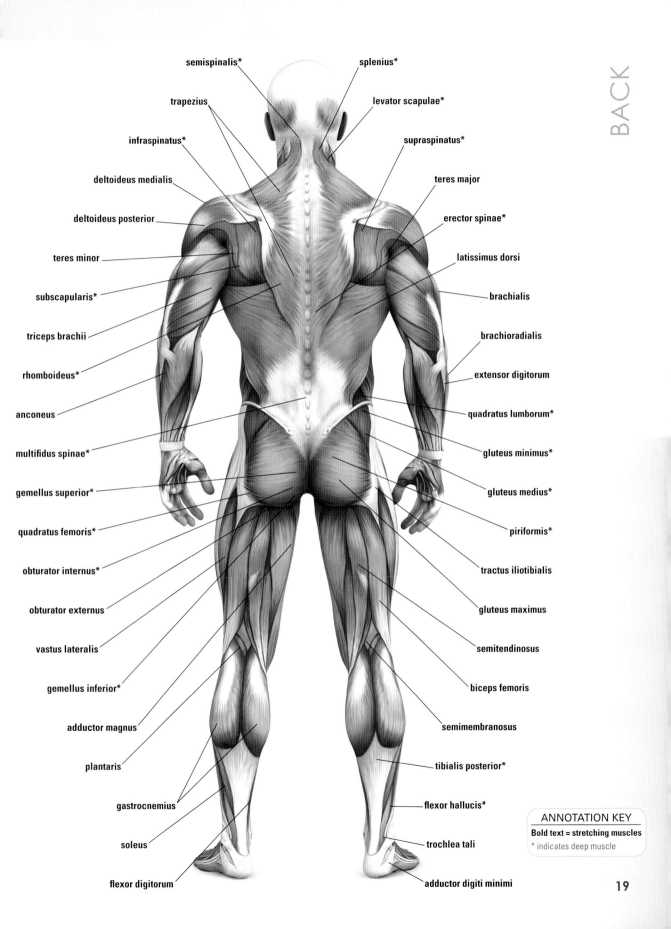

semispinalis*

splenius*

trapezius

levator scapulae*

infraspinatus*

supraspinatus*

deltoideus medialis

teres major

deltoideus posterior

erector spinae*

teres minor

latissimus dorsi

subscapularis*

brachialis

triceps brachii

brachioradialis

rhomboideus*

extensor digitorum

anconeus

quadratus lumborum*

multifidus spinae*

gluteus minimus*

gemellus superior*

gluteus medius*

quadratus femoris*

piriformis*

obturator internus*

tractus iliotibialis

obturator externus

gluteus maximus

vastus lateralis

semitendinosus

gemellus inferior*

biceps femoris

adductor magnus

semimembranosus

plantaris

tibialis posterior*

gastrocnemius

flexor hallucis*

soleus

trochlea tali

flexor digitorum

adductor digiti minimi

ANNOTATION KEY

Bold text = stretching muscles

* indicates deep muscle

19

THE STRETCHING SESSION

Like a dance number that moves with a natural grace and flow, so should your stretching routine. But grace and flow only come with practice. Practicing the following routine will benefit you—it will become easier with repetition because it's designed to flow from stretch to stretch.

Practice the Stretching Session so that you can recall the order without thinking, which shortens the movements between the stretches and creates a connective circuit. With the flow of movement uninterrupted by unnecessary level changes, it allows you to enter into a calm meditative state. You work on the mind as well as the body.

Isolation and Control
When performing this stretch routine, focus on isolation and control. Isolating a muscle helps you understand how it works. You can give it greater control over the stretch, which allows you the time and focus necessary to change and gauge the intensity. For beginners, the fewer muscles you try to stretch at one time, the better.

Controlling your stretch with proper weight distribution, involving the hands and arms, and proper form are important to help you gauge the intensity of the stretch. As you learn to control the range of movement and intensity of the stretches, you will rely less and less on a trainer. The goal here is for you to become a responsible, self-sufficient stretcher.

Holding the Stretch
Unless otherwise stated, hold each stretch for 30 seconds, with a total of two sets per

stretch, and a 10-second break between sets. This keeps the routine flowing and consistent. Wear a watch, use a stopwatch, count aloud, or keep a close eye on the clock—use whatever form of keeping track of time that works best for you.

Don't Forget to Breathe

We all do it, and need it to survive—we breathe. Still, almost everyone restricts and tries to control their breathing during a workout or stretch session. Do not hold your breath while stretching. Breathing helps to mechanically remove lactic acid and other by-products of exercise. Proper breathing helps relax the body and increases blood flow to the organs. You want to use your breath to communicate to your muscles and to communicate the intensity of the stretch, without becoming overly aware of your breathing. Try this technique to learn how to let your breathing happen naturally:

Relax your jaw, letting your mouth open slightly. This will relax the muscles in the back of the neck and in the diaphragm, which will allow oxygen in to feed your muscles.

Even though your jaw is relaxed, breathe through your nose. This not only cleans the air, but also ensures proper temperature and humidity for oxygen transfer into the lungs.

Exhale . . . Inhale

Exhale as you move into the stretch. Once in the stretch, imagine that you are inhaling healthy, oxygenated, cool air directly into the target muscle. The exhaled breath is warm and removes toxins, taking any negative feelings or emotions out with it. Empower yourself with every breath.

ASSISTED FOOT STRETCHES

POINT

1 Sit on a mat or chair, and cross your right leg over the left so that your ankle rests on top of your left thigh.

2 Brace your right ankle with your right hand and grasp the front of your right foot with your left hand. Press down on the top of your foot, focusing the palm of your hand on the knuckles of your toes so that they point inward.

3 Switch legs, and repeat on the other side.

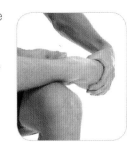

FLEXION

1 While still seated, again cross your right leg over the left so that your ankle rests on top of your thigh.

2 Brace your right heel with your right hand, and grasp the bottom of your toes and ball of the foot with your left hand.

3 Pull back on your toes until you feel a stretch in your arch.

4 Switch legs, and repeat on the other side.

TARGETS
- Feet
- Calves
- Arch of the foot

DO
- During the slope-down phase, push forcefully with your palms—their downward force must be stronger than the upward force of your pulling fingers.
- During the slope-up phase, push forcefully with your fingers—their upward force must be stronger than the downward force of your pushing palms.

AVOID
- Allowing your foot to shift—firmly stabilize your ankle and heel.

EXPERT'S TIP

Don't ignore your feet! They are the foundation for all standing movements. These stretches will strengthen your ankles and improve range of movement in the foot and calf for cardiovascular activities.

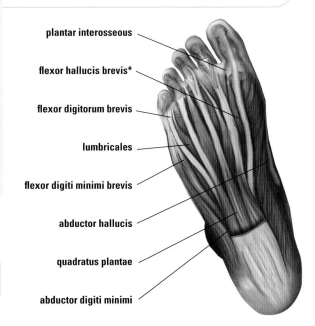

plantar interosseous

flexor hallucis brevis*

flexor digitorum brevis

lumbricales

flexor digiti minimi brevis

abductor hallucis

quadratus plantae

abductor digiti minimi

ANNOTATION KEY

Bold text = stretching muscles

* indicates deep muscle

SLOPE-DOWN

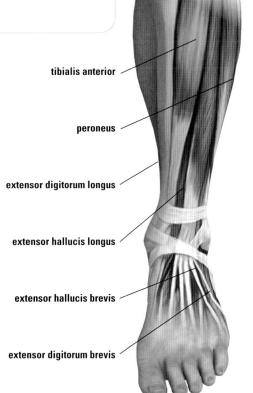

1 Sit on a mat or chair, and cross your right leg over the left so that your ankle rests on top of your left thigh.

2 Grasp your foot so that the palms of your hand lie across the top of your foot and your fingers are wrapped around the bottom.

3 Using your palms on the top outside of your foot, push down. At the same time, pull up the bottom of your foot with your fingers; this creates the "slope-down."

4 Switch legs, and repeat on the other side, before again crossing your right leg over the left so that your ankle rests on top of your thigh.

5 Grasp your foot so that the palms of your hand lie across the top of your foot and your fingers are wrapped around the bottom.

6 Using your palms on the top outside of your foot, push down. At the same time, pull up the bottom of your foot with your fingers; this creates the "slope-up."

7 Switch legs, and repeat on the other side.

SLOPE-UP

EXPERT'S TIP

Make sure that you don't "sickle" your foot, which is when your ankle turns in so that your big toe points in toward the other foot.

BEST FOR

- extensor digitorum longus
- extensor digitorum brevis
- tibialis anterior
- extensor hallucis longus
- extensor hallucis brevis
- flexor digitorum brevis
- quadratus plantae
- flexor digiti minimi brevis
- flexor hallucis brevis
- lumbricales
- plantar interosseous
- abductor hallucis
- abductor digiti minimi

ANNOTATION KEY
Bold text = stretching muscles

tibialis anterior

peroneus

extensor digitorum longus

extensor hallucis longus

extensor hallucis brevis

extensor digitorum brevis

BAND-ASSISTED STRETCHES

1. Sit on a chair with your feet flat on the floor.

2. Loop an elastic exercise band around the inside of your right foot, and then take hold of both ends of the band with your right hand.

3. Keeping your right leg stable, point your right foot and pull the band to the right, stretching the inside of the ankle.

4. Switch legs, and repeat on the other side.

TARGETS
• Ankles

DO
• Use a towel if you don't have an elastic exercise band.

AVOID
• Shifting your weight to the side you are stretching; your weight should be evenly balanced on your sitting bones.

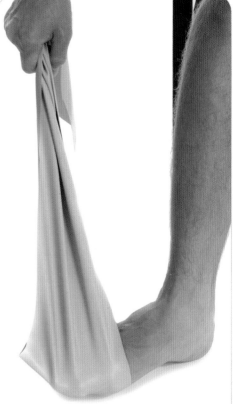

WING STRETCH

1 Sit on the floor with your legs extended in front of you. Bend both knees.

2 Loop an elastic exercise band around the inside of your right foot, and then take hold of both ends of the band with your left hand.

3 Keeping your right leg stable, point your right foot and pull the band to the left, stretching the outside of the ankle.

4 Switch legs, and repeat on the other side.

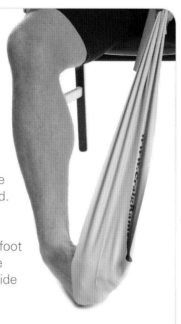

SICKLE STRETCH

BEST FOR

- peroneus longus
- peroneus brevis
- soleus
- gastrocnemius
- tibialis posterior

ANNOTATION KEY
Bold text = stretching muscles
* indicates deep muscle

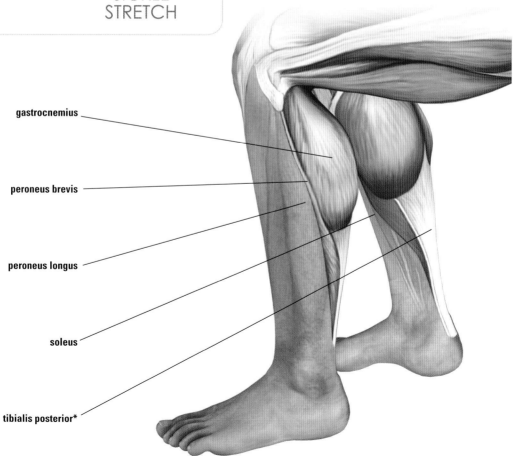

gastrocnemius

peroneus brevis

peroneus longus

soleus

tibialis posterior*

SEATED LEG CRADLE

1. Sit on the floor with your legs extended in front of you.

2. Bend your right knee and grasp your calf with your right hand. With your left hand, support the raised foot as you hug it into your chest as if you were cradling a baby. Keep your heel roughly 12 inches away from your chest.

3. Switch sides, and repeat on the other leg.

TARGETS
- Upper hamstrings
- Gluteal region

DO
- Keep your chest lifted.
- Contract your gluteal muscles.

AVOID
- Holding your breath.

ANNOTATION KEY
Bold text = stretching muscles
* indicates deep muscle

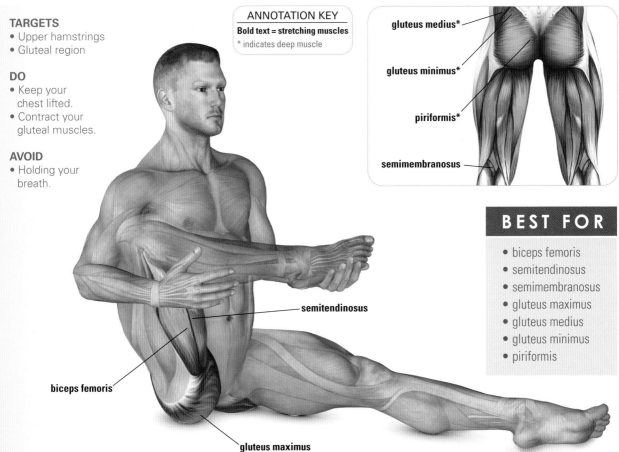

gluteus medius*

gluteus minimus*

piriformis*

semimembranosus

semitendinosus

biceps femoris

gluteus maximus

BEST FOR

- biceps femoris
- semitendinosus
- semimembranosus
- gluteus maximus
- gluteus medius
- gluteus minimus
- piriformis

UNILATERAL SEATED FORWARD BEND

1. Sit on the floor, sitting up as straight as possible, with your legs extended in front of you in parallel position.

2. Bend your left leg until it is turned out, with the bottom of your left foot resting at your right inner thigh just above the kneecap. Rest your hands on your knee.

3. Bend from your waist, and lean forward over your right leg. Place your forearms above your right kneecap.

4. Switch legs, and repeat on the other side.

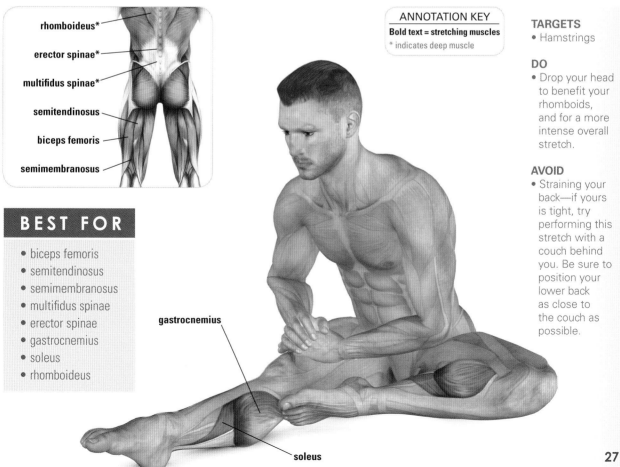

- rhomboideus*
- erector spinae*
- multifidus spinae*
- semitendinosus
- biceps femoris
- semimembranosus

gastrocnemius

soleus

ANNOTATION KEY
Bold text = stretching muscles
* indicates deep muscle

TARGETS
- Hamstrings

DO
- Drop your head to benefit your rhomboids, and for a more intense overall stretch.

AVOID
- Straining your back—if yours is tight, try performing this stretch with a couch behind you. Be sure to position your lower back as close to the couch as possible.

BEST FOR

- biceps femoris
- semitendinosus
- semimembranosus
- multifidus spinae
- erector spinae
- gastrocnemius
- soleus
- rhomboideus

27

BILATERAL SEATED FORWARD BEND

1. Sit on the floor, sitting up as straight as possible with your back flattened and your legs extended in front of you in parallel position. Your feet should be relaxed and flexed slightly.

2. Lean forward, lowering your abdominals over your thighs, forearms resting above your kneecaps as you stretch.

3. Slowly roll up, and repeat if desired.

TARGETS
• Hamstrings

DO
• Bend at the hips and keep your spine fairly straight as you stretch.
• Extend your torso as far forward over your legs as possible.

AVOID
• Holding your breath.

Modification

Advanced: For a deeper stretch in your hamstrings, place an elastic exercise band around the balls of your feet, using both hands to draw the band upward.

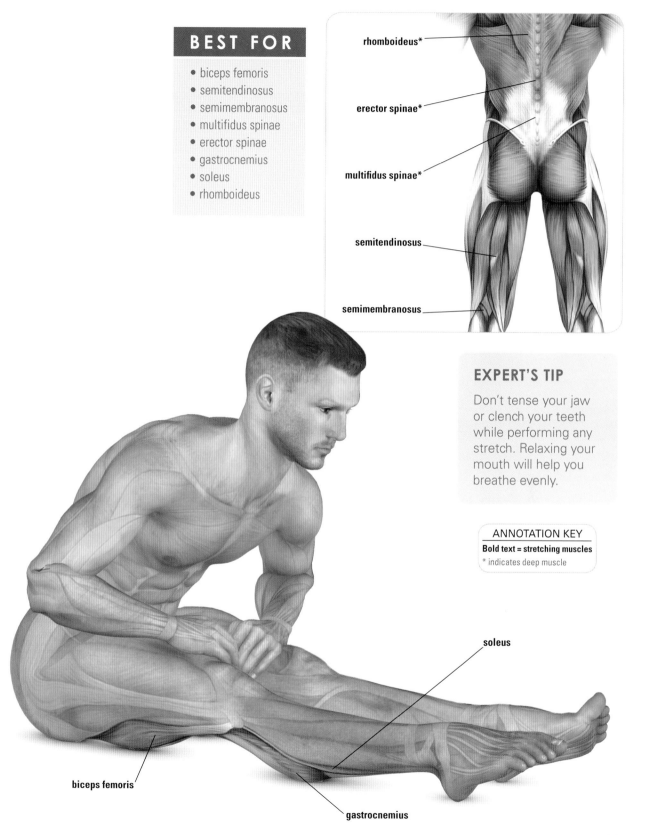

BEST FOR

- biceps femoris
- semitendinosus
- semimembranosus
- multifidus spinae
- erector spinae
- gastrocnemius
- soleus
- rhomboideus

rhomboideus*

erector spinae*

multifidus spinae*

semitendinosus

semimembranosus

EXPERT'S TIP

Don't tense your jaw or clench your teeth while performing any stretch. Relaxing your mouth will help you breathe evenly.

ANNOTATION KEY

Bold text = stretching muscles
* indicates deep muscle

soleus

biceps femoris

gastrocnemius

BUTTERFLY STRETCHES

SEATED BUTTERFLY

1 Sit up tall on the floor or a mat, with the soles of your feet pressed together.

2 Place your forearms or elbows on your inner thighs, and grab your feet and toes with your hands.

3 Draw your heels in toward your core.

TARGETS
- Adductors
- Lower back
- Trunk extensors

DO
- Exhale as you drop your chest toward the floor.

AVOID
- Slouching.
- Holding your breath.
- Rocking backward, off your hip bones; instead, feel them anchored on the floor.

EXPERT'S TIP

Do not compromise the position of your upper body: sit nice and tall, and try to feel your hip bones on the floor.

BEST FOR
- adductor longus
- adductor magnus
- adductor brevis
- gracilis
- pectineus
- obturator externus
- erector spinae
- quadratus lumborum

FOLDED BUTTERFLY

1 From the Seated Butterfly, place your forearms or elbows on your inner thighs, and grab your feet and toes with your hands. Keep your heels a comfortable distance from your core.

2 Fold your upper body forward until you feel a stretch in your groin and in your upper inner thighs.

3 Slowly roll up, and repeat if desired.

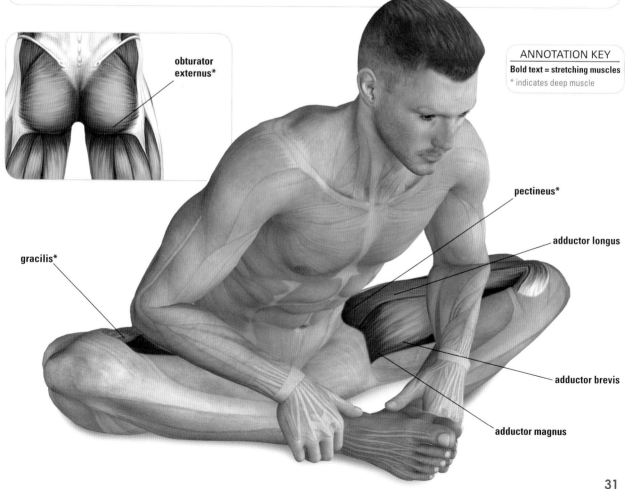

obturator externus*

ANNOTATION KEY
Bold text = stretching muscles
* indicates deep muscle

pectineus*

adductor longus

gracilis*

adductor brevis

adductor magnus

SCOOP RHOMBOIDS

1. Sit on the floor and extend your legs in front of you in parallel position. Bend your knees slightly, keeping your heels on the floor.

2. Grasp beneath your hamstrings with your hands.

3. Keeping your chin down, round your upper back down as you lean back toward the floor. Hold for 10 to 15 seconds.

4. Slowly roll up to the starting position, and repeat if desired.

TARGETS
• Upper back

DO
• Exhale as you round your upper back and lean backward.

AVOID
• Holding your breath.

BEST FOR

• rhomboideus

EXPERT'S TIP

This stretch will improve mobility in the back muscles and reduce tension.

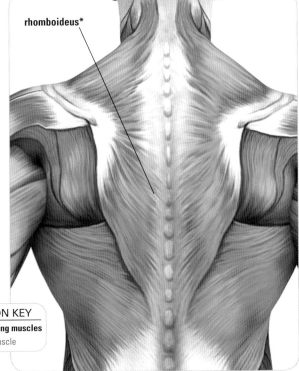

rhomboideus*

ANNOTATION KEY
Bold text = stretching muscles
* indicates deep muscle

FRONT DELTOID TOWEL STRETCH

BEST FOR

- deltoideus anterior

1 Sit on the floor with your legs extended in parallel position, knees slightly bent and heels on the floor. Grip a small towel behind your back, palms facing behind you.

2 Gently slide your buttocks forward along the floor until you feel a comfortable stretch in your front deltoids. Return to the starting position, and repeat if desired.

TARGETS
- Shoulders

DO
- Keep your hands together while gripping the towel.

AVOID
- Leaning your head forward; instead, keep it in line with your body.

EXPERT'S TIP

This stretch improves mobility in people who regularly perform repetitive shoulder and arm motions, such as hairdressers and cashiers.

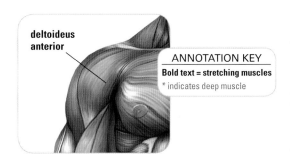

deltoideus anterior

ANNOTATION KEY

Bold text = stretching muscles
* indicates deep muscle

LYING-DOWN ARCH STRETCH

1 Lie flat on your back with your legs extended in front of you, toes pointed.

2 Extend your arms on the floor above your head, biceps to your ears and fingers pointed, lengthening your entire body as you stretch.

TARGETS
- Abdominals
- Intercostal muscles
- Middle back

DO
- Form one long line from your fingertips to your toes.

AVOID
- Overarching your lower back.

BEST FOR

- rectus abdominis
- transversus abdominis
- latissimus dorsi
- intercostales interni
- intercostales interni

ANNOTATION KEY
Bold text = stretching muscles
* indicates deep muscle

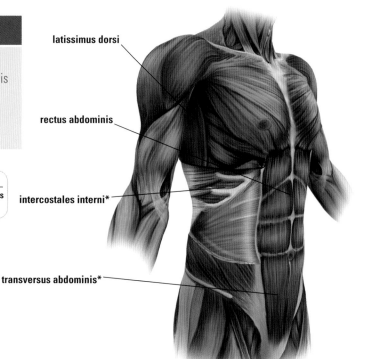

latissimus dorsi

rectus abdominis

intercostales interni*

transversus abdominis*

LYING-DOWN GROIN STRETCH

① Lie flat on your back with your legs extended in front of you. Turning out your legs from your hips, touch the soles of your feet together.

② Bend your knees and draw your feet in along the floor toward your body, keeping the soles together.

③ Place your hands on your inner thighs.

BEST FOR

- adductor longus
- adductor magnus
- adductor brevis
- gracilis
- pectineus
- obturator externus

ANNOTATION KEY
Bold text = stretching muscles
* indicates deep muscle

TARGETS
- Groin muscles

DO
- To deepen the stretch, keep the upper outside of your feet on the floor and lift your heels upward.

AVOID
- Lifting your lower back off the floor.
- Bouncing your legs open to achieve a deeper stretch.

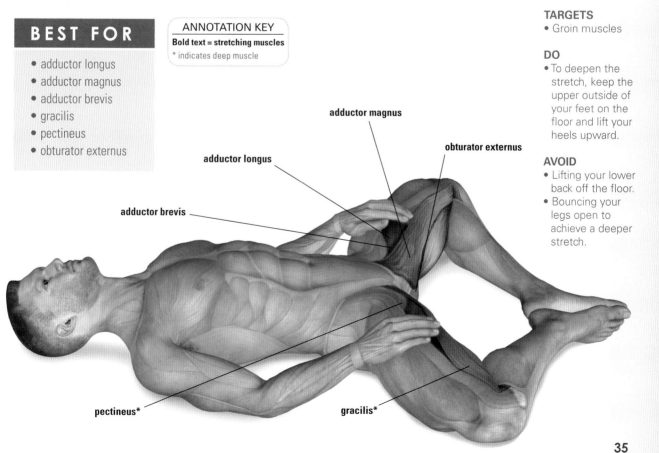

adductor magnus

obturator externus

adductor longus

adductor brevis

pectineus*

gracilis*

LYING-DOWN PRETZEL STRETCH

1 Lie on your back, with both legs elongated and parallel and your arms extended away from your torso, palms facing up.

2 Bend your right leg, placing the sole of your foot on the floor.

TARGETS
- Rotator muscles
- Gluteal region
- Chest

DO
- Keep your elbows and wrists lower than your shoulders, protecting your rotator cuff.

AVOID
- Lifting your shoulders; try to keep both shoulder blades in contact with the floor throughout the stretch.

3 Carefully lift your buttocks off the floor, tilting your torso 2 to 3 inches to your left, and cross your right leg over to your left side, with your knee bent at a right angle.

4 Hold, return to the starting position, and repeat on the other side.

BEST FOR

- gemellus inferior
- gemellus superior
- gluteus medius
- gluteus minimus
- piriformis
- obturator externus
- obturator internus
- pectoralis major
- pectoralis minor
- quadratus femoris
- gluteus maximus

EXPERT'S TIP

Before you cross one leg over the other, ensure that your body is in a straight line from your head to your elongated, pointed toe.

Modification

Advanced: Place the palm of your left hand on your right quadriceps while your right leg is crossed over your left, and vice versa.

ANNOTATION KEY

Bold text = stretching muscles
* indicates deep muscle

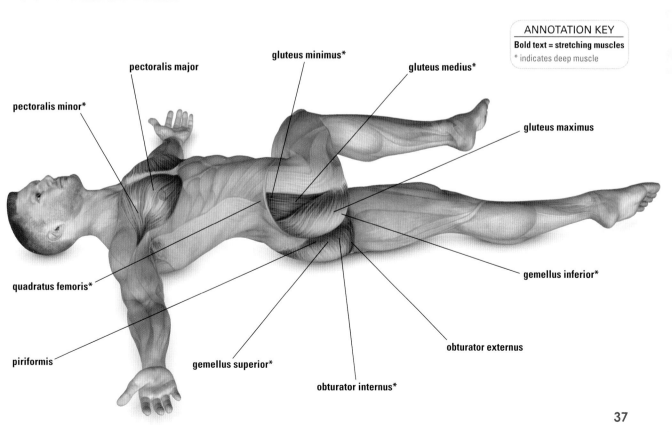

gluteus minimus*

pectoralis major

gluteus medius*

pectoralis minor*

gluteus maximus

quadratus femoris*

gemellus inferior*

piriformis

gemellus superior*

obturator internus*

obturator externus

37

UNILATERAL LEG STRETCHES

UNILATERAL KNEE-TO-CHEST STRETCH

1 Lie on your back, and bend your right knee in toward your chest.

2 Placing your hands on your right hamstrings, gently hug your knee closer to your chest as you stretch.

TARGETS
- Lower back
- Groin muscles
- Gluteal region
- Hamstrings

DO
- Keep your lower back on the floor—tucking your pelvis just ¼ inch will help keep your back grounded.

AVOID
- Lifting your head or upper back.
- Holding your breath.

EXPERT'S TIP

Perform both of the Unilateral Leg Stretches followed by the Hip Adductor Stretch (see pages 40–41), as a smooth sequence on one side before switching to the other and performing all three stretches on the opposite side.

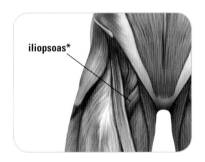

iliopsoas*

Modifications

Advanced: Place your hands on the top of your shin where it connects with your knee before hugging the knee into your chest.

Extreme: Bring your bent leg toward your chest to form a right angle and hold it at the ankle, using your hands to draw the knee even closer to your chest, intensifying the stretch in your hamstrings and iliopsoas.

UNILATERAL LEG RAISE

1 With your hands placed on your hamstrings just below the knee, extend and straighten your right leg toward the ceiling.

2 Point both feet.

3 Switch your hand position, right hand on your right calf muscle, left hand on your hamstring. Gently bring your thigh toward your chest, increasing the intensity of the stretch.

4 Prepare to move into the Hip Adductor Stretch (see pages 40–41).

Modification

Advanced: Wrap a towel or an elastic exercise band around the ball of your foot, and then gently pull both ends to increase the flex, intensifying the stretch.

BEST FOR

- erector spinae
- gluteus maximus
- gluteus medius
- gluteus minimus
- biceps femoris
- semitendinosus
- semimembranosus
- iliopsoas
- gastrocnemius
- soleus

ANNOTATION KEY

Bold text = stretching muscles

* indicates deep muscle

semimembranosus

semitendinosus

biceps femoris

soleus

erector spinae*

gluteus medius*

gluteus minimus*

gluteus maximus

gastrocnemius

39

HIP ADDUCTOR STRETCH

1 Continuing from the leg-raised position of the Unilateral Leg Raise (see page 39), release your left hand from your right hamstrings, and lower your left arm to the floor.

TARGETS
- Hip adductors

DO
- Flex the foot of your extended leg to ensure a deep stretch.

AVOID
- Lifting your lower back off the floor.
- Holding your breath.

ANNOTATION KEY

Bold text = stretching muscles

* indicates deep muscle

EXPERT'S TIP

When lowering your arm to the floor in step 2, keep your elbow and wrist slightly lower than your shoulder to protect your rotator cuff.

BEST FOR

- adductor longus
- adductor magnus
- adductor brevis
- gracilis
- pectineus
- obturator externus
- biceps femoris
- semitendinosus
- semimembranosus

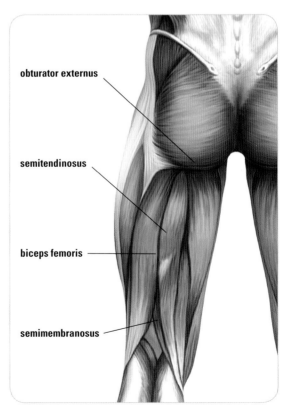

obturator externus

semitendinosus

biceps femoris

semimembranosus

2 Keeping your right hand on your right calf muscle, extend the straightened leg out to the side of the body, and point your foot.

3 Release, and repeat the Unilateral Knee-to-Chest Stretch, the Unilateral Leg Raise, and the Hip Adductor Stretch on the opposite side.

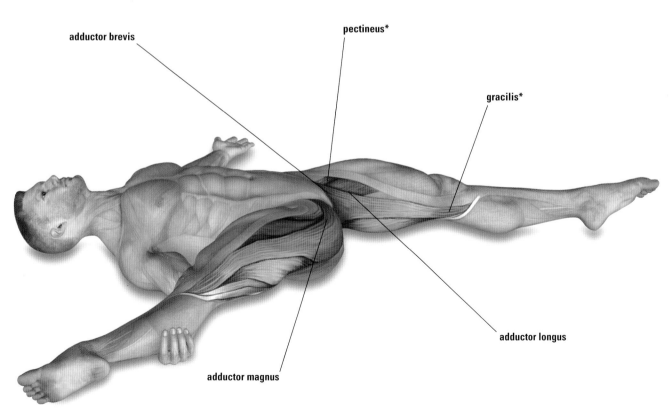

adductor brevis

pectineus*

gracilis*

adductor longus

adductor magnus

LYING-DOWN FIGURE 4

1 Lie on your back with your legs extended.

2 Point both toes. Bend your right knee and turn the leg out so that your right ankle rests on your left thigh just above the knee, creating a figure 4.

TARGETS
• Gluteal region

DO
• Keep your head and shoulder blades on the floor.

AVOID
• Twisting your lower body; instead, keep your hips square.

BEST FOR

• gluteus maximus
• gluteus medius
• gluteus minimus
• piriformis

3 Bend your left leg, drawing both legs (still in the figure 4 position) in toward your chest as you grasp the back of your left thigh.

4 Push your right elbow against your right inner thigh, turning out the right leg slightly to increase the intensity of the stretch.

5 Return to the starting position, switch legs, and repeat.

ANNOTATION KEY

Bold text = stretching muscles
* indicates deep muscle

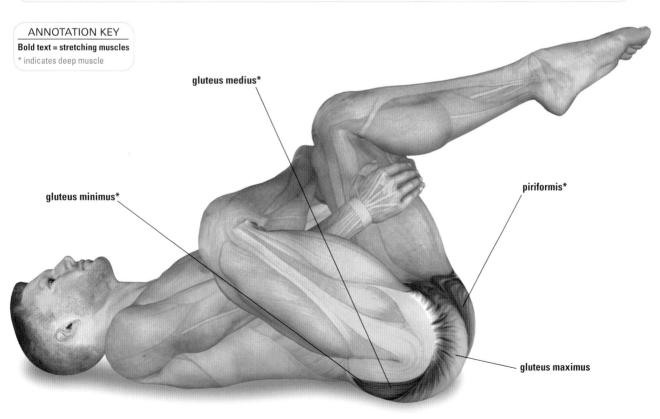

gluteus medius*

gluteus minimus*

piriformis*

gluteus maximus

INTERNAL HIP ROTATOR STRETCH

❶ Lie on your back with your arms extended at your sides.

❷ Bend your knees, planting your feet generously outside of shoulder-width apart.

TARGETS
• Hip rotators

DO
• Keep your abdominals tight and rest your hands on the floor to support your lower back.

AVOID
• Lifting your lower back and glutes.

❸ Keeping the rest of your body still, rotate your right hip inward, bringing your knee toward the floor.

EXPERT'S TIP

This stretch is not about big moves: internally rotate your hip just 2 to 5 inches as you stretch.

4 Slowly return to the starting position, and repeat with the opposite leg.

BEST FOR

- gluteus medius
- gluteus minimus
- tensor fasciae latae

ANNOTATION KEY

Bold text = stretching muscles
* indicates deep muscle

tensor fasciae latae

gluteus minimus*

gluteus medius*

HAPPY BABY STRETCH

1 Lie on your back.

2 Bend your knees in toward your chest and grasp the outsides of your feet with your hands.

TARGETS
- Gluteal region
- Hamstrings
- Lower back

DO
- Keep your elbows slightly bent.
- Draw your shoulders toward the floor.
- Tuck your pelvis forward about ¼ inch to engage your abdominals and keep your lower back anchored.

AVOID
- Lifting your head or shoulder blades off the floor.

BEST FOR

- gluteus maximus
- gluteus medius
- gluteus minimus
- piriformis
- biceps femoris
- semitendinosus
- semimembranosus
- erector spinae
- multifidus spinae

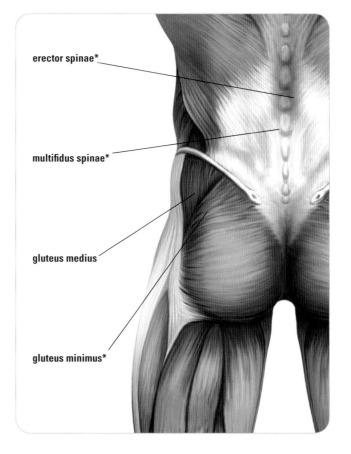

erector spinae*

multifidus spinae*

gluteus medius

gluteus minimus*

3 Bring your knees toward the floor as you open your legs.

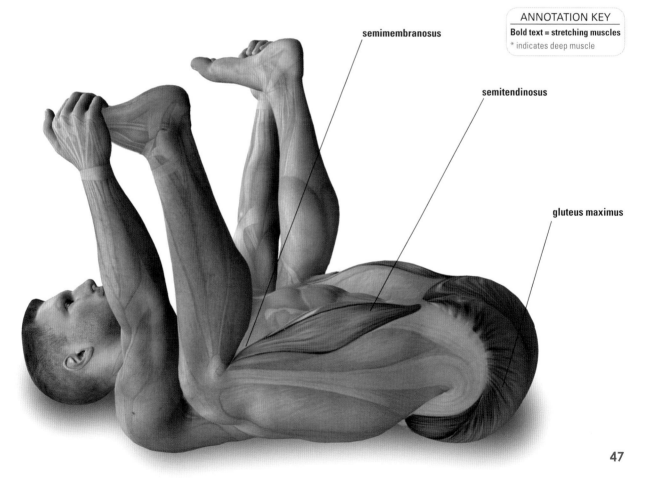

semimembranosus

semitendinosus

gluteus maximus

47

SIDE-LYING RIB STRETCH

❶ Lie on your right side with your legs together and extended. Place both palms on the floor, your right arm supporting you and your left arm positioned in front of your body. Your upper body should be slightly lifted.

❷ Bend your left leg and rest the foot just in front of your right thigh, knee pointing up toward the ceiling.

TARGETS
- Rib cage
- Obliques
- Outer thighs
- Lower back

DO
- Shift your weight forward on your supporting hip.
- Place a towel under your bottom hip if it feels uncomfortable to rest directly on the floor.

AVOID
- Tightening your jaw, which can cause tension in your neck.

❸ Keeping your legs in place, press down with your hands and straighten both arms as you raise your body upward, feeling a stretch around your right rib cage.

❹ Release, switch sides, and repeat.

BEST FOR

- obliquus externus
- obliquus internus
- tensor fasciae latae
- multifidus spinae
- erector spinae

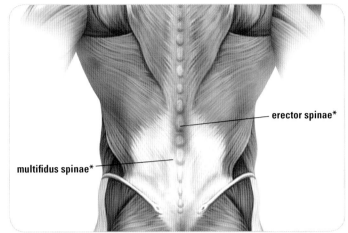

erector spinae*

multifidus spinae*

ANNOTATION KEY
Bold text = stretching muscles
* indicates deep muscle

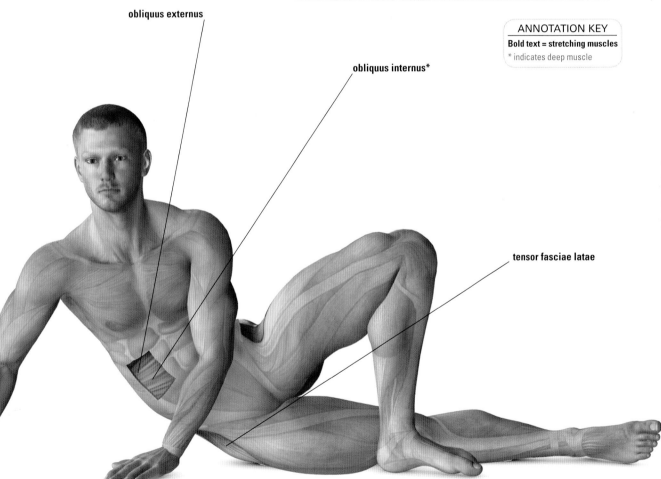

obliquus externus

obliquus internus*

tensor fasciae latae

SIDE-LYING KNEE BEND

1 Lie on your left side, with your legs extended together in line with your body. Extend your left arm, and rest your head on your upper arm.

TARGETS
• Quadriceps

DO
• Keep your knees together, one on top of the other.
• Tuck your pelvis slightly forward and lift your chest to engage and stretch your core.
• Keep your foot pointed and parallel with your leg.

AVOID
• Leaning back onto your gluteal muscles.

2 Bend your right knee and grasp the ankle with your right hand.

3 Pull your ankle in toward your buttocks as you stretch.

4 Return to the starting position, and repeat on the other side.

EXPERT'S TIP

Place a towel under your bottom hip if it feels uncomfortable to rest directly on the floor.

BEST FOR

- rectus femoris
- vastus lateralis
- vastus intermedius
- vastus medialis

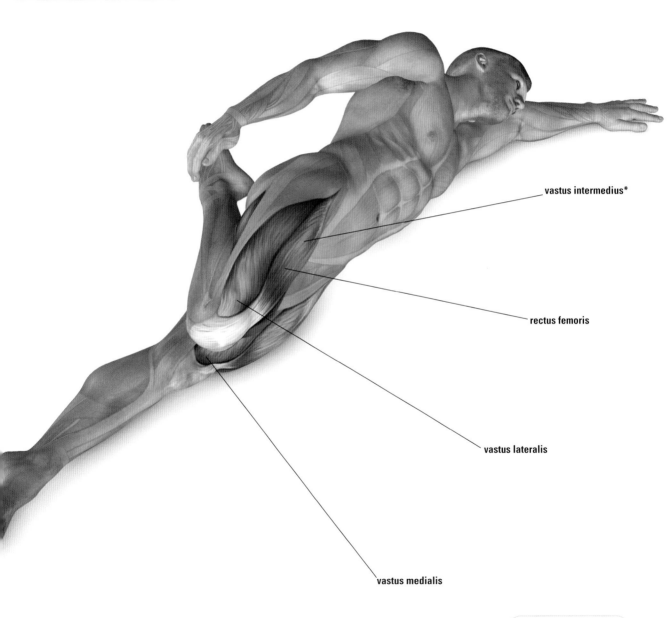

vastus intermedius*

rectus femoris

vastus lateralis

vastus medialis

ANNOTATION KEY

Bold text = stretching muscles
* indicates deep muscle

COBRA STRETCH

1 Lie facedown, legs extended behind you with toes pointed. Position the palms of your hands on the floor slightly above your shoulders, and rest your elbows on the floor.

2 Push down into the floor, and slowly lift through the top of your chest as you straighten your arms.

3 Pull your tailbone down toward your pubis as you push your shoulders down and back.

4 Elongate your neck and gaze forward.

TARGETS
• Abdominals

DO
• Maintain pressure between the floor and your hips.
• Relax your shoulders, and keep them down and away from your ears.

AVOID
• Tipping your head too far backward.
• Overdoing this stretch—it can lead to excessive pressure on your lower back.

BEST FOR

• rectus abdominis
• transversus abdominis
• obliquus externus
• obliquus internus

EXPERT'S TIP

Feel your chest moving forward as well as upward; this will reduce the risk of straining your lower back.

Modification

Easier: Follow steps 1 and 2, only rising to rest on your forearms.

ANNOTATION KEY

Bold text = stretching muscles

* indicates deep muscle

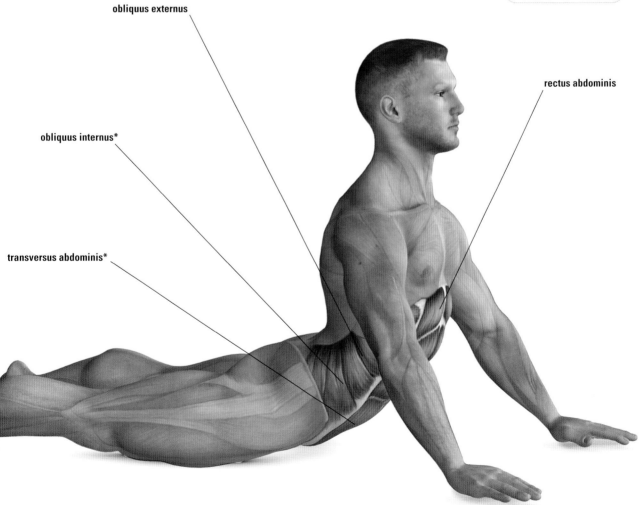

obliquus externus

rectus abdominis

obliquus internus*

transversus abdominis*

BACK STRETCHES

CHILD'S POSE

❶ Begin in the Cobra Stretch (see pages 52–53), with your legs extended behind you and your palms and elbows resting on the floor.

❷ Bend your knees, straighten your arms, and press down into the floor to draw your body back onto your knees. Sit back on your heels.

❸ Drop your torso down onto your thighs, and extend your arms on the floor above your head so that your forehead comes to rest on the floor.

TARGETS
- Back
- Quadriceps

DO
- During the Child's Pose, rest your forehead on a towel or mat if desired.

AVOID
- Tensing your neck and shoulders.
- Hyperextending your lower back or arms.
- Holding your breath.

ANNOTATION KEY
Bold text = stretching muscles
* indicates deep muscle

EXPERT'S TIP
The Child's Pose is a relaxing stretch that you can perform to relieve the tension of stressful situations.

BEST FOR
- rectus femoris
- vastus lateralis
- vastus intermedius
- vastus medialis
- multifidus spinae
- erector spinae
- rhomboideus
- trapezius

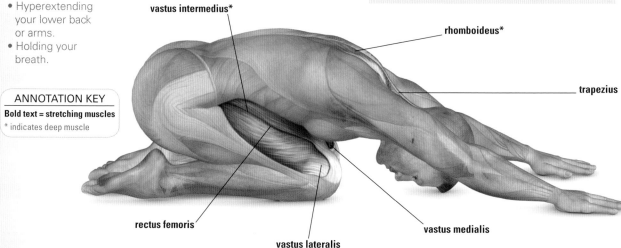

vastus intermedius*

rhomboideus*

trapezius

rectus femoris

vastus lateralis

vastus medialis

KNEELING LAT STRETCH

1 Begin in the Child's Pose—kneeling so that your buttocks are above your knees, your torso is leaning forward, and your arms are extended in front of you, palms down. Carefully rest your forehead on the floor.

2 Bend your left arm so that it is perpendicular to your torso, keeping your palm flat on the floor.

3 Return to the starting position, and repeat on the other side.

BEST FOR

• latissimus dorsi

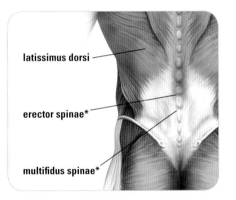

latissimus dorsi

erector spinae*

multifidus spinae*

ANNOTATION KEY

Bold text = stretching muscles

* indicates deep muscle

EXPERT'S TIP

During the Cat Stretch, push down with your hands and knees to achieve maximum contraction.

CAT STRETCH

1 From the Kneeling Lat Stretch, rise up to rest on all fours, with your hands at shoulder width and your knees spread 2 to 3 inches apart.

2 Round your spine upward as you draw your navel in toward your spine, keeping your hips lifted and your shoulders stable.

3 Hold the stretch at the top, and release.

BEST FOR

• erector spinae

PIGEON STRETCH

1 Kneel with your buttocks resting lightly on your heels and your arms at your sides, supporting some of your weight.

EXPERT'S TIP

For an even more advanced stretch, lean your torso forward until your head is resting on your crossed forearms.

BEST FOR

- adductor longus
- adductor magnus
- adductor brevis
- gracilis
- pectineus
- obturator externus
- rectus femoris
- vastus lateralis
- vastus intermedius
- vastus medialis
- biceps femoris
- semitendinosus
- semimembranosus
- gluteus maximus
- gluteus medius
- gluteus minimus
- iliopsoas

TARGETS
- Gluteal area
- Groin muscles
- Hamstrings
- Quadriceps

DO
- Maintain a slight bend in your elbows.
- Lean primarily on your bent leg.

AVOID
- Hyperextending your elbows.

2 Straighten your left leg to extend it along the floor behind you, keeping the leg in parallel, aligned with your body, including your right knee, which should be facing straight forward.

3 Move your arms forward to rest slightly in front of your right knee. Your hands should be shoulder-width apart and flat on the floor, palms down.

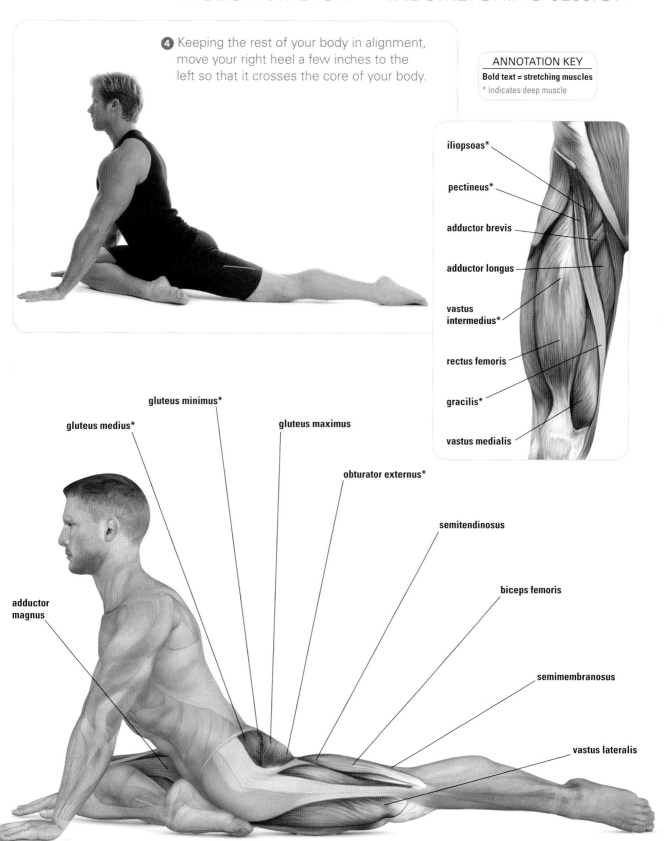

❹ Keeping the rest of your body in alignment, move your right heel a few inches to the left so that it crosses the core of your body.

ANNOTATION KEY

Bold text = stretching muscles
* indicates deep muscle

iliopsoas*

pectineus*

adductor brevis

adductor longus

vastus intermedius*

rectus femoris

gracilis*

vastus medialis

gluteus minimus*

gluteus medius*

gluteus maximus

obturator externus*

semitendinosus

biceps femoris

adductor magnus

semimembranosus

vastus lateralis

SHIN STRETCH

① Kneel with your buttocks resting lightly on your heels.

BEST FOR

- gastrocnemius
- soleus
- rectus femoris
- vastus lateralis
- vastus intermedius
- vastus medialis

TARGETS
- Shins
- Quadriceps

DO
- Contract and engage your gluteal muscles to avoid a curve in your lumbar spine. You will have space between the heels and the glutes.

AVOID
- Arching your back.

② Place your hands flat on the floor behind you, with your fingers pointing forward. Keep a slight bend in your elbows.

③ Lean back slightly to increase the intensity of the stretch.

ANNOTATION KEY

Bold text = stretching muscles

* indicates deep muscle

vastus intermedius*

rectus femoris

vastus medialis

vastus lateralis

gastrocnemius

soleus

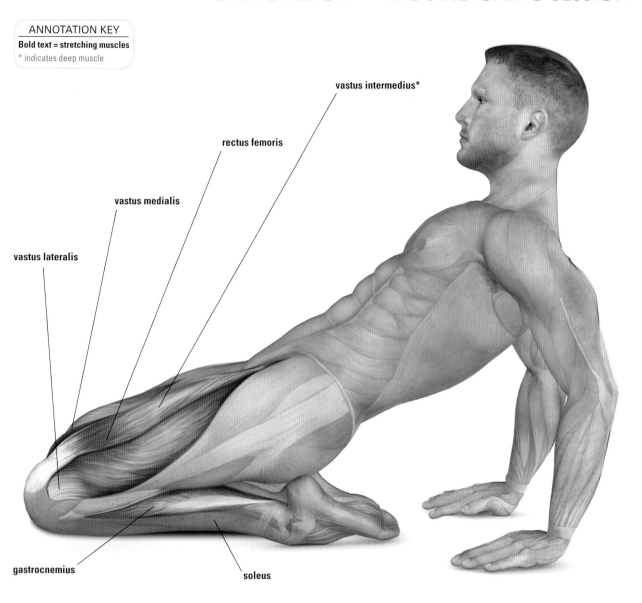

Modification

Advanced: Carefully bend your elbows until they rest on the floor. If more of a stretch is still required, lower down farther until your shoulder blades rest against the floor and your arms straighten beside your calves.

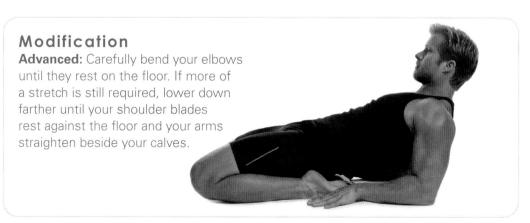

FROG STRADDLE

1 Kneel on all fours.

2 Bend your elbows and shift your weight forward so that you are leaning onto your elbows and forearms.

TARGETS
• Inner thighs
• Hip adductors

DO
• Stretch until you reach a point of challenge without feeling pain.

AVOID
• Placing too much of your weight on your kneecaps.
• Allowing your lower back to sink.

3 Spread your knees apart, drawing your feet in slightly and putting some weight on them to take pressure off your kneecaps.

4 Lower your legs and buttocks down to the floor and bring the soles of your feet together to deepen the stretch.

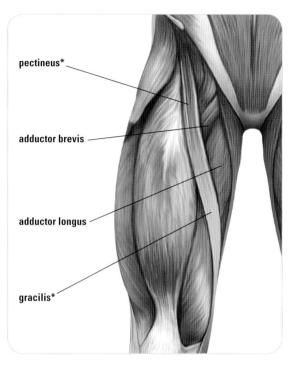

pectineus*

adductor brevis

adductor longus

gracilis*

Modification

Advanced: Move your forearms forward and lean into them. Try to keep both your pelvis and your heels on the floor as you stretch.

ANNOTATION KEY
Bold text = stretching muscles
* indicates deep muscle

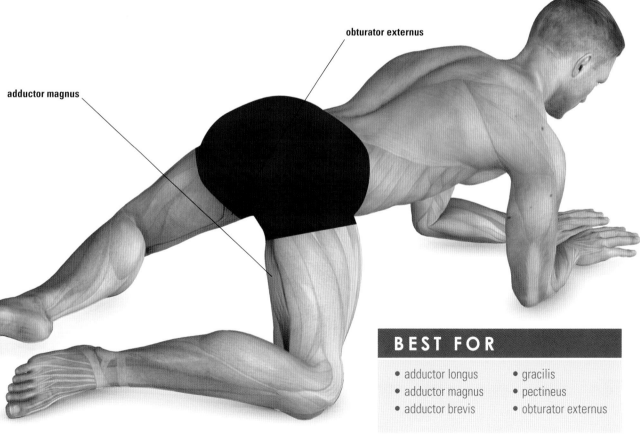

obturator externus

adductor magnus

BEST FOR

- adductor longus
- gracilis
- adductor magnus
- pectineus
- adductor brevis
- obturator externus

61

HALF STRADDLE STRETCHES

HALF STRADDLE

1 Sit upright with your legs in front of you, knees bent.

2 Keeping your right knee bent, open it up so that it touches the floor.

3 Extend your left leg straight out to the side of your body.

4 Plant your arms on the floor behind you to support your lower back as you stretch.

TARGETS
- Hamstrings
- Quadriceps
- Inner thighs
- Calves
- Obliques

DO
- Lean against a couch to stabilize yourself and to correctly align your hip bones on the floor.
- During the side-leaning portion, lean from your hips, elongating your upper torso and reaching for your lower thigh and kneecap.

AVOID
- Lifting your buttocks from the floor.

ANNOTATION KEY
Bold text = stretching muscles
* indicates deep muscle

EXPERT'S TIP
Perform both of the Half Straddle Stretches as a smooth sequence on one side before switching to the other.

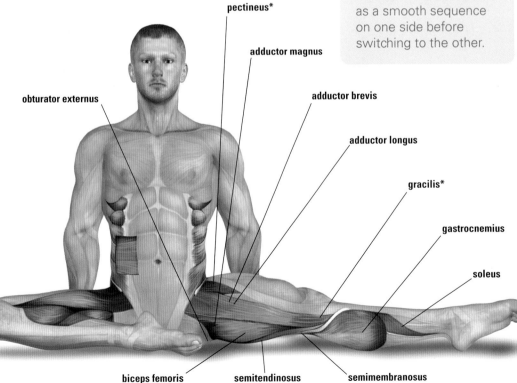

pectineus*

adductor magnus

obturator externus

adductor brevis

adductor longus

gracilis*

gastrocnemius

soleus

biceps femoris

semitendinosus

semimembranosus

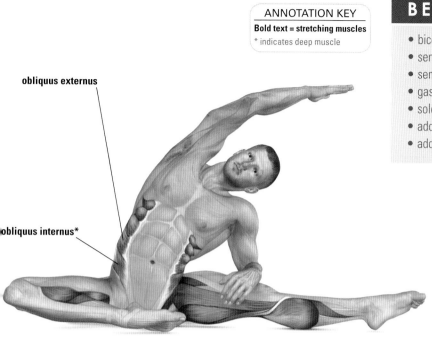

ANNOTATION KEY
Bold text = stretching muscles
* indicates deep muscle

obliquus externus

obliquus internus*

BEST FOR

- biceps femoris
- semitendinosus
- semimembranosus
- gastrocnemius
- soleus
- adductor longus
- adductor magnus
- adductor brevis
- gracilis
- pectineus
- obturator externus
- obliquus externus
- obliquus internus

EXPERT'S TIP

To focus the stretch on the upper portion of your hamstrings, point the toes of the outstretched leg downward. To emphasize the lower hamstrings, point your toes upward.

SIDE-LEANING HALF STRADDLE

❶ Begin in the Half Straddle Stretch: seated upright, with one leg bent on the floor and your other leg extended to the side. Bend your elbow, resting your forearm over your thigh.

❷ Raise your arm over your head, palm inward.

❸ Slowly bend slightly forward at the hips and lean toward your side until you feel a comfortable stretch.

❹ Return to the center, and perform both the Half Straddle and the Side-Leaning Half Straddle on the other side.

Modification
Advanced: Lower your elbow and forearm to the floor in front of your inner thigh.

DOUBLE-LEG STRADDLE SPLIT

1 Sit upright, with your legs extended and turned out from the hips as far you can comfortably reach. Your feet should be slightly flexed so that your toes point upward.

2 Place one hand on the floor directly in front of you and the other directly behind you. Align your hip bones on the floor.

3 Switch your hands, and repeat.

TARGETS
- Hip adductors
- Hamstrings

DO
- Sit up as tall as possible, with your torso long.

AVOID
- Leaning back as you stretch.
- Rounding your lower back.
- Bouncing your legs more widely open—you want to feel the stretch but never pain.

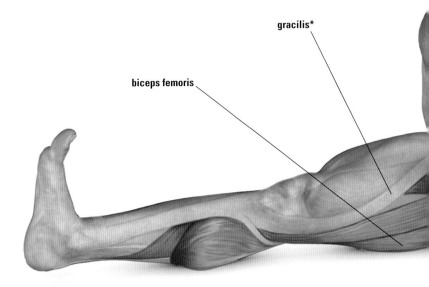

gracilis*

biceps femoris

EXPERT'S TIP

To take pressure off the lumbar region of your back, sit up against a couch or other stable object.

BEST FOR

- biceps femoris
- semitendinosus
- semimembranosus
- adductor longus
- adductor magnus
- adductor brevis
- gracilis
- pectineus
- obturator externus

Modification

Advanced: Press your hands down into the floor, slightly lifting your body upward. Carefully move your pelvis forward, increasing the stretch, and then lower yourself back to the floor and point your toes. Repeat three times, holding each stretch for 20 to 30 seconds.

pectineus*

adductor brevis

adductor longus

adductor magnus

ANNOTATION KEY

Bold text = stretching muscles
* indicates deep muscle

semimembranosus

obturator externus

semitendinosus

CHEST-TO-THIGH STRADDLE SPLIT

1 Begin in the Double-Leg Straddle Split (see page 64), sitting upright with your legs extended and turned out from the hips as far you can comfortably reach. Your feet should be slightly flexed so that your toes point upward.

TARGETS
- Inner thighs
- Hamstrings
- Gluteal area
- Rib cage
- Hip adductors

DO
- Keep the turnout in your legs, and aim your toes straight up.

AVOID
- Lifting your hip bones off the floor.
- Allowing your legs to shift inward.

2 Twist and lean your torso over your right thigh.

3 With your hands on the floor on either side of your right leg, reach your chest toward your thigh.

4 Return to the center, and repeat on the other side.

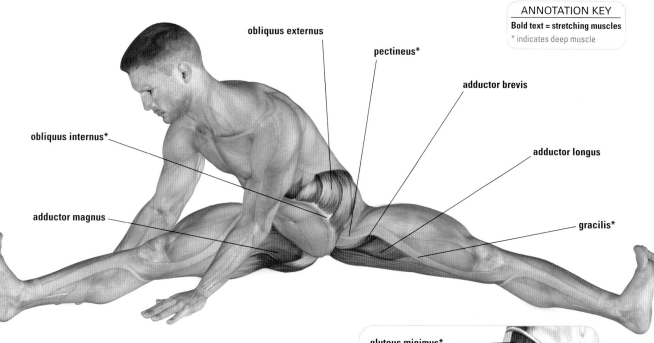

obliquus externus

pectineus*

adductor brevis

adductor longus

obliquus internus*

adductor magnus

gracilis*

ANNOTATION KEY
Bold text = stretching muscles
* indicates deep muscle

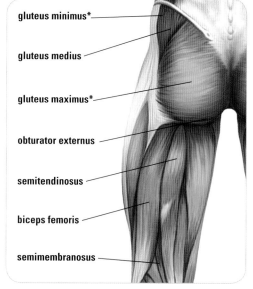

gluteus minimus*

gluteus medius

gluteus maximus*

obturator externus

semitendinosus

biceps femoris

semimembranosus

EXPERT'S TIP

Carefully push the back of the kneecap slightly down toward the floor, decreasing the amount of space between the back of the knee and the floor.

BEST FOR

- gluteus maximus
- gluteus medius
- gluteus minimus
- biceps femoris
- semitendinosus
- semimembranosus
- obliquus externus
- obliquus internus
- adductor longus
- adductor magnus
- adductor brevis
- gracilis
- pectineus
- obturator externus

CHEST-TO-FLOOR STRADDLE SPLIT

❶ Begin in the Double-Leg Straddle Split (see page 64), sitting upright with your legs extended and turned out from the hips as far as you can comfortably reach. Your feet should be slightly flexed so that your toes point upward.

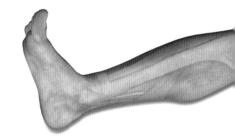

BEST FOR

- gluteus maximus
- gluteus medius
- gluteus minimus
- biceps femoris
- semitendinosus
- semimembranosus
- obliquus externus
- obliquus internus
- adductor longus
- adductor magnus
- adductor brevis
- erector spinae
- multifidus spinae
- obturator externus

TARGETS
- Hamstrings
- Lower back
- Hip adductors
- Gluteal area

DO
- Keep your torso long and flat, and your chest lifted.
- Lengthen your torso as you bend forward.
- Keep your legs turned out from the hips.

AVOID
- Overdoing this stretch—move carefully and slowly.

❷ Place your hands in front of your body and "walk" them forward as you bring your upper body toward the floor.

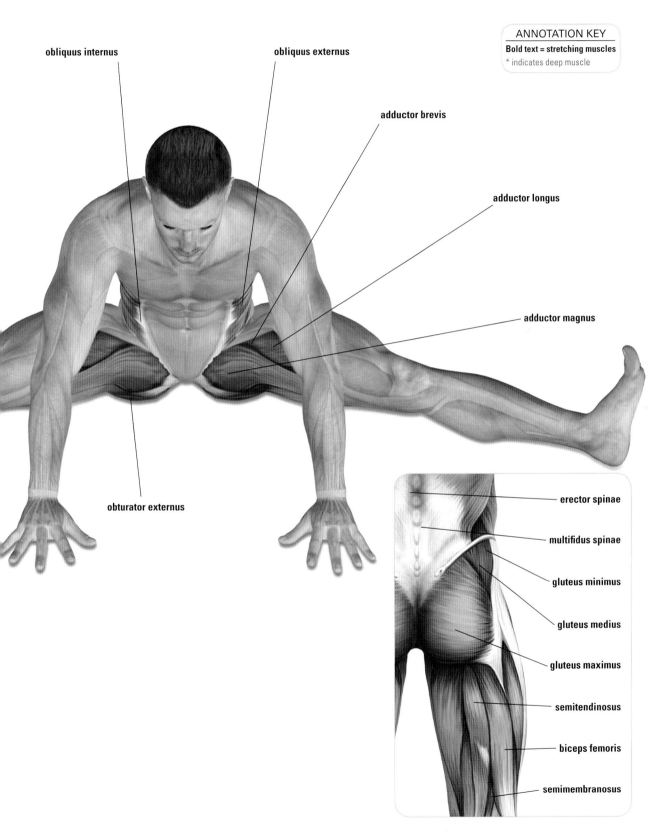

ANNOTATION KEY
Bold text = stretching muscles
* indicates deep muscle

obliquus internus

obliquus externus

adductor brevis

adductor longus

adductor magnus

obturator externus

erector spinae

multifidus spinae

gluteus minimus

gluteus medius

gluteus maximus

semitendinosus

biceps femoris

semimembranosus

TOE TOUCH

1 Stand with your legs and feet parallel and shoulder-width apart. Bend your knees very slightly.

2 Slowly round your spine downward, from your neck through your lower back, and lower your arms down the sides of your legs.

3 Continue to lower downward as you bend at the waist and let the weight of your body draw your head toward the floor as you stretch.

EXPERT'S TIP

To release from this stretch, follow right into the Standing Back Roll (see page 71).

TARGETS
- Hamstrings
- Upper back
- Lower back
- Calves

DO
- Relax your neck and jaw.
- Breathe naturally and steadily throughout the stretch.

AVOID
- Allowing your knees to touch— keeping your thighs spread slightly apart will help your gluteal muscles to engage and stretch.

BEST FOR
- biceps femoris
- semitendinosus
- semimembranosus
- rhomboideus
- erector spinae
- gastrocnemius
- soleus

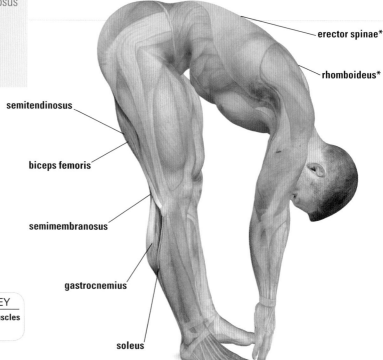

erector spinae*

rhomboideus*

semitendinosus

biceps femoris

semimembranosus

gastrocnemius

soleus

ANNOTATION KEY
Bold text = stretching muscles
* indicates deep muscle

STANDING BACK ROLL

① From the bottom position of the Toe Touch, slowly roll up halfway to the point at which you feel your gluteal muscles above your hips and thighs.

② Cross your forearms to place your hands on the opposite thighs, and round your shoulders forward.

③ Feel the heaviness of your head as you stretch your upper back between the shoulder blades.

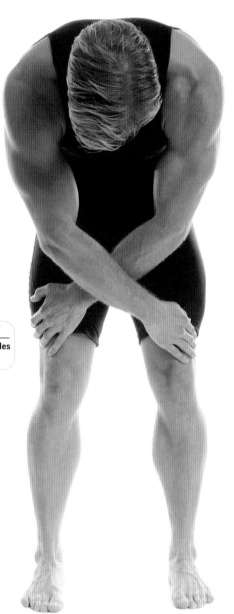

BEST FOR

• rhomboideus

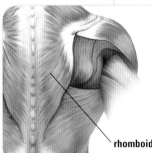

ANNOTATION KEY
Bold text = stretching muscles
* indicates deep muscle

rhomboideus*

EXPERT'S TIP

Imagine a sense of "contraction," as if someone has punched your upper stomach in an upward motion.

TARGETS
• Upper back
• Middle back

DO
• Keep your knees slightly bent.
• Tuck your pelvis forward slightly, allowing your upper body to "contract."

AVOID
• Allowing your knees to turn inward.

GOOD MORNING STRETCH

① Stand with your legs and feet parallel and shoulder-width apart. Bend your knees very slightly.

② Tuck your pelvis about ¼ inch forward.

TARGETS
- Back
- Neck
- Abdominals
- Obliques
- Palms
- Forearms
- Upper arms

DO
- Keep your elbows slightly bent.
- Tuck your pelvis.

AVOID
- Hyperextending either your lower back or elbows.

EXPERT'S TIP

This invigorating stretch is a great one to perform first thing in the morning.

③ Reach your arms all the way up toward the ceiling, keeping them long and in parallel with your body. Focus your energy on the middle of your palms, which should be facing inward. Turn your gaze upward as you stretch.

BEST FOR

- trapezius
- biceps brachii
- levator scapulae
- scalenus
- splenius
- sternocloidomastoideus
- extensor carpi radialis longus
- extensor carpi radialis brevis
- extensor carpi ulnaris
- flexor carpi radialis
- flexor carpi ulnaris
- rhomboideus
- rectus abdominis
- brachialis
- brachioradialis
- transversus abdominis
- obliquus externus
- obliquus internus
- latissimus dorsi
- palmaris longus

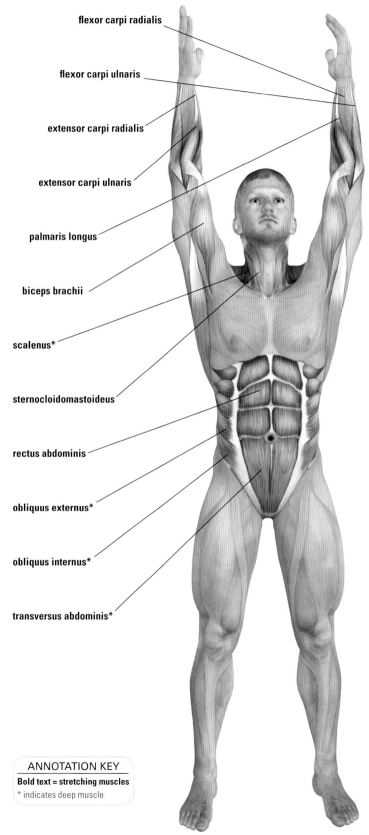

flexor carpi radialis

flexor carpi ulnaris

extensor carpi radialis

extensor carpi ulnaris

palmaris longus

biceps brachii

scalenus*

sternocloidomastoideus

rectus abdominis

obliquus externus*

obliquus internus*

transversus abdominis*

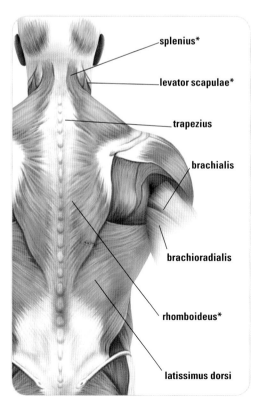

splenius*

levator scapulae*

trapezius

brachialis

brachioradialis

rhomboideus*

latissimus dorsi

ANNOTATION KEY

Bold text = stretching muscles
* indicates deep muscle

SCALP AND FACIAL STRETCHES

SCALP STRETCH

1. Place your palms on your temples. Slide your fingers open and then slightly back along your scalp.

2. Mainly utilizing your thumbs, grasp a handful of hair on either side of your head.

3. Gently pull your hair away from your head until you feel a slight tension on your scalp.

EXPERT'S TIP

Augment the stretch by using the tips of your fingers to gently pull on your hair (if you have enough hair to do so). And remember: do not tug too hard. Give your scalp a gentle stretch without pulling your hair out.

BEST FOR

- frontalis
- occipitalis
- galea aponeurotica

TARGETS
- Scalp
- Facial muscles

DO
- Keep your head steady throughout the stretches.

AVOID
- Tensing your neck and shoulders.

ANNOTATION KEY
Bold text = stretching muscles
gray text = fibrous tissue
* indicates deep muscle

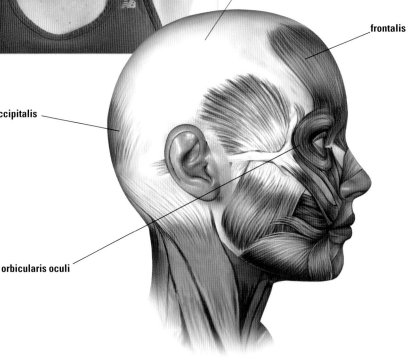

galea aponeurotica

frontalis

occipitalis

orbicularis oculi

LION STRETCH

① Raise both eyebrows as you try to lift your ears upward and backward.

② Open your mouth wide, and place the tip of your tongue behind the back of your bottom teeth, flexing your tongue out as much as comfortably possible.

③ Flare your nostrils, and hold for 5 seconds.

④ Release the stretch, and then repeat for a total of three sets, holding each for 5 seconds.

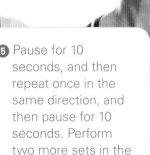

frontalis

corrugator

masseter

pterygoideus

lingua

orbicularis orbis

BEST FOR

- orbicularis orbis
- temporalis
- masseter
- pterygoideus internus
- corrugator
- lingua
- frontalis

EYE BOX STRETCH

① Gaze up toward the upper right corner of your field of vision. Hold for 3 seconds.

② Move your focus clockwise to the lower right corner of your field of vision. Hold for 3 seconds.

③ Move your focus to the lower left corner of your field of vision. Hold for 3 seconds.

BEST FOR

- orbicularis oculi

④ Move your focus to the to the upper left corner of your field of vision. Hold for 3 seconds.

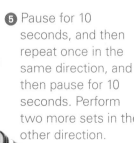

⑤ Pause for 10 seconds, and then repeat once in the same direction, and then pause for 10 seconds. Perform two more sets in the other direction.

NECK STRETCHES

SIDE NECK TILT

1. Stand with your legs and feet parallel and shoulder-width apart. Bend your knees very slightly.

2. Tuck your pelvis about ¼ inch forward and stand tall, with your chest slightly lifted and shoulders pressed lightly downward and back, away from your ears.

3. Slowly tilt your head to the right, feeling the weight of your head shifting in this direction as you hold for 5 seconds.

4. Slowly return your head to the center, rest for 5 seconds, and repeat on the other side.

BEST FOR

- levator scapulae

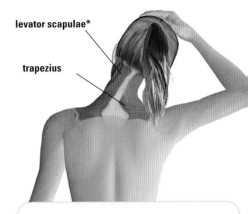

levator scapulae*

trapezius

Modification
Advanced: Place the palm of one hand over your head with your fingertips touching your ear. Stretch your other arm downward, and extend your fingertips as if you were trying to grab something just out of reach. Dancers call this "reaching for the keys."

TARGETS
- Neck

DO
- Breathe easily and normally during all of the stretches.

AVOID
- Lifting or tensing your shoulders.

DOWNWARD NECK TILT

1. Still standing, look down, focusing your nose toward your armpit.

2. Hold for 5 seconds, return back to the neutral position, rest for 5 seconds, and repeat to the right for a second set.

BEST FOR

- trapezius

Modification
Advanced: To deepen this stretch, again place one hand on your head, and "reach for the keys" with the other.

UPWARD NECK TILT

1 Still standing, slowly tilt your head so that your nose points toward the upper left. Focus your gaze upward.

2 Hold for 5 seconds.

3 Slowly return your head to the center, rest for 5 seconds, and repeat on the other side.

BEST FOR

- sternocleidomastoid

Modification

Advanced: To deepen this stretch, again place one hand on your head, and "reach for the keys" with the other.

NECK AND HEAD TURN

1 Drop your hand and then lift your chin very slightly and focus straight ahead.

2 Turn your head to the right side, and hold for 5 seconds.

3 Slowly return your head to the center, rest for 5 seconds, and repeat on the other side.

BEST FOR

- sternocleidomastoideus
- splenius
- levator scapulae
- trapezius
- ligamentum interspinalis
- ligamentum capsular facet

BACK-OF-THE-NECK STRETCH

BEST FOR

- sternocleidomastoideus
- ligamentum nuchae
- ligamentum supraspinous
- trapezius

1 Clasp your hands behind your head, interlacing your fingers. Gently tilt your head forward, and hold for 5 seconds.

2 Slowly bring your head back up, rest for 5 seconds, and repeat.

ANNOTATION KEY

Bold text = stretching muscles
italic text = ligaments
* indicates deep muscle

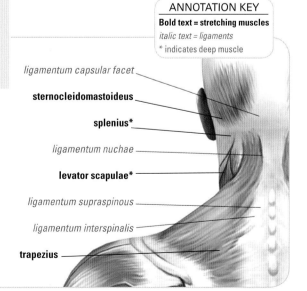

ligamentum capsular facet

sternocleidomastoideus

splenius*

ligamentum nuchae

levator scapulae*

ligamentum supraspinous

ligamentum interspinalis

trapezius

77

TRICEPS STRETCH

1. Stand with your legs and feet parallel and shoulder-width apart. Bend your knees very slightly, and tuck your pelvis slightly forward, lift your chest, and press your shoulders downward and back.

2. Reach your right arm up behind your head, and bend it from the elbow, aiming to bring your elbow toward the middle of the back of your head. Your right hand should fall between your shoulder blades.

3. Grab your right elbow with your left hand, and gently pull to intensify the stretch while the elbow stays still.

4. Release your elbow, and repeat on the other side.

TARGETS
• Upper arms

DO
• Keep your shoulders pressed down and back, away from your ears.
• Maintain a firm, stable midsection, keeping your pelvis slightly tucked.

AVOID
• Tilting your head and/or neck forward, jeopardizing your spinal alignment.
• Holding your breath.

triceps brachii

BEST FOR

• triceps brachii

ANNOTATION KEY
Bold text = stretching muscles
* indicates deep muscle

BICEPS STRETCH

① Stand with your legs and feet parallel and shoulder-width apart. Bend your knees very slightly, and tuck your pelvis slightly forward, lift your chest, and press your shoulders downward and back.

② Clasp your hands together behind your back with your palms together, straighten your arms, and twist your wrists inward, bringing your palms to your gluteal muscles.

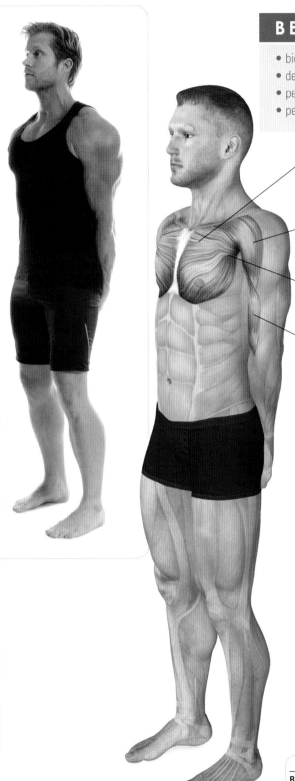

BEST FOR

- biceps brachii
- deltoideus anterior
- pectoralis major
- pectoralis minor

pectoralis major

deltoideus anterior

pectoralis minor*

biceps brachii

TARGETS
- Upper arms
- Shoulders
- Chest

DO
- Keep your shoulders pressed down and back, away from your ears.

AVOID
- Collapsing your chest forward.

ANNOTATION KEY

Bold text = stretching muscles

* indicates deep muscle

WALL-ASSISTED CHEST STRETCH

❶ Stand parallel to a wall, with the wall on the left side of your body.

BEST FOR

- pectoralis minor
- deltoideus anterior
- pectoralis major

❷ Extend your left arm back against the wall, so that your palm is flat against it.

TARGETS
- Chest
- Shoulders

DO
- Keep your shoulders pressed down and back, away from your ears.
- Position the arm against the wall so that your elbow is slightly lower than your shoulder and your wrist is slightly below your elbow, at a slight diagonal, to protect your rotator cuff from injury.

AVOID
- Rotating your chest and/or torso toward the wall when lunging; instead, face forward.

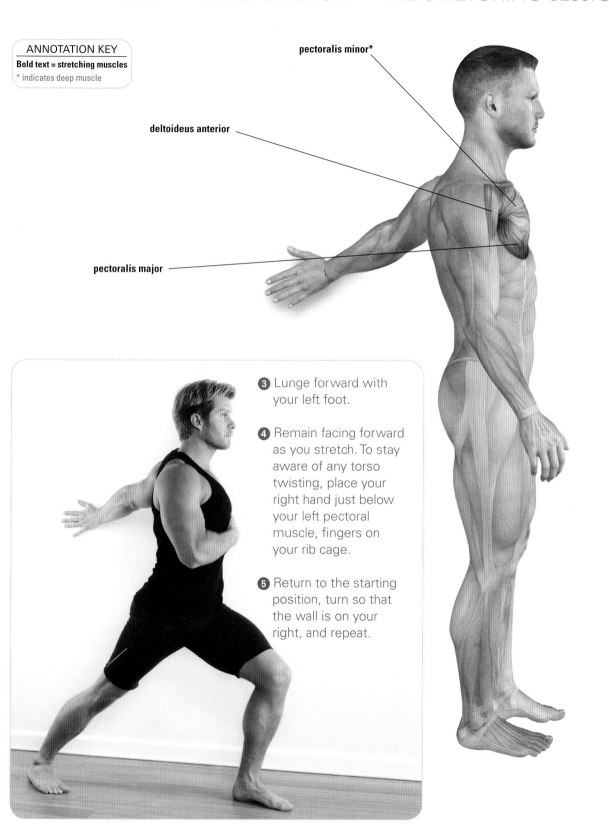

ANNOTATION KEY
Bold text = stretching muscles
* indicates deep muscle

pectoralis minor*

deltoideus anterior

pectoralis major

❸ Lunge forward with your left foot.

❹ Remain facing forward as you stretch. To stay aware of any torso twisting, place your right hand just below your left pectoral muscle, fingers on your rib cage.

❺ Return to the starting position, turn so that the wall is on your right, and repeat.

FOREARM STRETCHES

WRIST FLEXION

BEST FOR

- extensor carpi radialis
- extensor carpi ulnaris
- extensor digiti minimi
- extensor digitorum
- extensor indicis
- extensor pollicis

TARGETS
- Wrists
- Hands
- Forearms

DO
- Keep in mind that a flexing movement stretches the extensor muscles, and an extending movement stretches the flexors.
- Be sure to press your thumb into the meaty part of your palm, attached to the thumb, intensifying the stretch in your forearm and wrist.

AVOID
- Lifting or tensing your shoulders.

① Stand or sit with your arms at your sides.

② Bend your right forearm up from the elbow, creating a 90-degree bend. Your palm should be facing the floor.

③ Drop and flex your right wrist downward so that your palm faces inward.

④ Place your left fingers over the back of your right hand, and your left thumb on the palm of the hand, directly on the right thumb muscle.

⑤ Gently press your left fingers into the back of your right hand, bringing your right wrist to a 60- to 90-degree bend, while pressing your left thumb into the palm away from the body, creating a deeper stretch.

⑥ Release, switch hands, and repeat on the other side.

EXPERT'S TIP

Perform these stretches after a long phone conversation or a stressful commute to release any tension in your hands and forearms, or if you often carry weight in your arms, such as when holding a child.

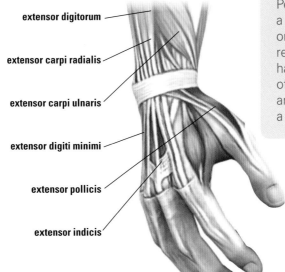

extensor digitorum

extensor carpi radialis

extensor carpi ulnaris

extensor digiti minimi

extensor pollicis

extensor indicis

ANNOTATION KEY

Bold text = stretching muscles

* indicates deep muscle

WRIST EXTENSION

BEST FOR

- flexor carpi radialis
- flexor carpi ulnaris
- flexor digiti minimi
- flexor digitorum
- palmaris longus
- flexor pollicis

1 Stand or sit with your arms at your sides.

2 Bend your right forearm up from the elbow, creating a 90-degree bend. Your palm should be facing up toward the ceiling.

3 Drop and flex your right wrist downward so that your palm faces outward.

4 Place your left fingers over the back your right hand, and your left thumb on the palm of the hand, directly on the right thumb muscle.

5 Using your left thumb and palm, gently press the right thumb and palm in toward your body. At the same time, use your left fingers to press in on the back of the right hand, thus flattening the right palm and creating a deeper stretch.

EXPERT'S TIP

Imagine that you are holding the eraser end of a pencil under each arm. Engage the muscles around your armpits to hold onto the imaginary pencil—keeping your shoulders perfectly positioned in the process. Use this technique for all stretches and resistance training that involve holding your elbows in toward your rib cage.

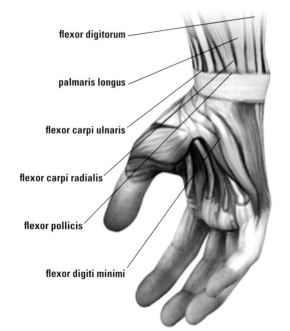

flexor digitorum

palmaris longus

flexor carpi ulnaris

flexor carpi radialis

flexor pollicis

flexor digiti minimi

ANNOTATION KEY

Bold text = stretching muscles
* indicates deep muscle

CALF STRETCHES

CALF HEEL DROP

① Stand on a step, a raise, or a stair with your legs and feet parallel and shoulder-width apart. Bend your knees very slightly and tuck your pelvis slightly forward, lift your chest, and press your shoulders downward and back.

② Position your left foot slightly in front of your right, and place the ball of your right foot on the edge of the step.

BEST FOR

- gastrocnemius
- soleus
- tendo calcaneus

③ Drop your right heel down while controlling the amount of weight on the right leg to increase or decrease the intensity of the stretch in the right calf.

④ Release, switch feet, and repeat on the other side.

TARGETS
- Calves
- Achilles tendon

DO
- Use the wall or other stable object to balance yourself if necessary.
- Engage each head of your calf muscles by gently and slowly rolling from your big toe to your small toe and back again, shifting your body weight over your toes as you go.

AVOID
- Bouncing to achieve a greater stretch—all of your movements should be preformed slowly and carefully.

EXPERT'S TIP

Perform a few calf stretches during your cardiovascular workout—this will help relieve calf tightness and stress throughout your workout.

TOE-UP CALF STRETCH

1 Stand with your legs and feet parallel and shoulder-width apart. Bend your knees very slightly and tuck your pelvis slightly forward, lift your chest, and press your shoulders downward and back.

EXPERT'S TIP

Avoid focusing downward— this might take the necessary weight off the front foot and onto the back leg, greatly decreasing the intensity of the stretch.

2 Position the ball of your right foot on a step or against a wall.

3 With your knees straight, bring your hips forward.

4 Release, switch feet, and repeat on the other side.

gastrocnemius

soleus

tendo calcaneus

STANDING QUADRICEPS STRETCH

BEST FOR

- rectus femoris
- vastus lateralis
- vastus intermedius
- vastus medialis
- tibialis anterior
- extensor digitorum brevi

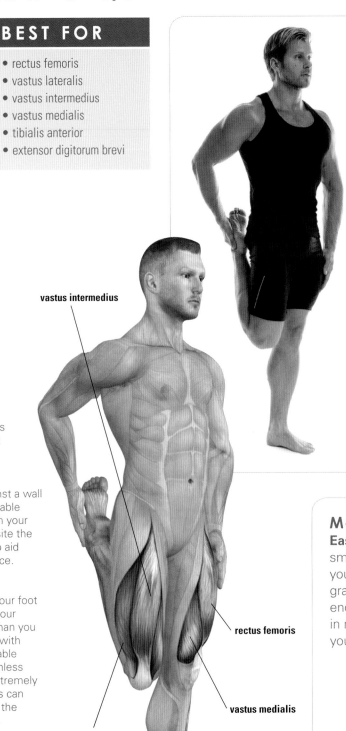

vastus intermedius

TARGETS
- Quadriceps
- Top of foot
- Ankles

DO
- Lean against a wall or other stable object with your arm opposite the bent leg to aid your balance.

AVOID
- Bringing your foot closer to your buttocks than you can reach with a comfortable stretch. Unless you are extremely limber, this can compress the knee joint.

rectus femoris

vastus medialis

vastus lateralis

tibialis anterior

extensor digitorum brevi

ANNOTATION KEY
Bold text = stretching muscles
* indicates deep muscle

❶ Stand with your legs and feet parallel and shoulder-width apart. Bend your knees very slightly and tuck your pelvis slightly forward, lift your chest, and press your shoulders downward and back.

❷ Bend your right knee behind you so that your ankle is raised toward your buttocks.

❸ Reach down with your right hand to grab your foot just below your ankle and gently pull as you stretch.

❹ Release the stretch, switch legs, and repeat.

Modification

Easier: Wrap a small towel around your ankle and grasp both ends to aid in raising your foot.

KNEELING SPRINTER STRETCH

1. Kneel with your left knee bent, your toe pointed and extended behind you on the floor. Bend your right leg so that your right foot is flat on the floor, next to your left knee.

2. Position your hands on the floor just beyond shoulder-width apart, slightly in front of your body, with palms downward.

3. Sit back on your left heel as you lean slightly forward.

4. Release the stretch, switch legs, and repeat.

BEST FOR

- soleus
- extensor digitorum brevis
- tendo calcaneus

ANNOTATION KEY
Bold text = stretching muscles
Italic text = ligaments
* indicates deep muscle

TARGETS
- Calves
- Achilles tendon

DO
- Keep the sole of your front foot and the upper arch of your back foot on the floor.
- Lean your chest farther forward over your raised upper leg to increase the intensity of the stretch.

AVOID
- Allowing your foot to roll inward.

tendo calcaneus

extensor digitorum brevis

soleus

SUMO SQUAT

1 Stand with your feet planted beyond shoulder width and your toes turned out. Bend your knees very slightly and tuck your pelvis slightly forward, lift your chest, and press your shoulders downward and back.

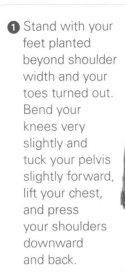

2 Place your hands on your thighs.

3 Squat down until your thighs are parallel to the ground, keeping your weight on your heels.

4 Push off with your heels at the bottom of the squat, squeezing your glutes and inner thighs to rise back to the starting position.

TARGETS
- Groin muscles
- Hip adductors

DO
- "Sit" rather than bend your legs to squat: this prevents straining your knees.
- Tuck your pelvis and lift your chest throughout the stretch.
- Use your hands on your thighs to slightly open up your thighs, helping you obtain proper turnout from the hips and achieve perfect form.

AVOID
- Letting your knees extend beyond your toes.
- Hunching your shoulders.

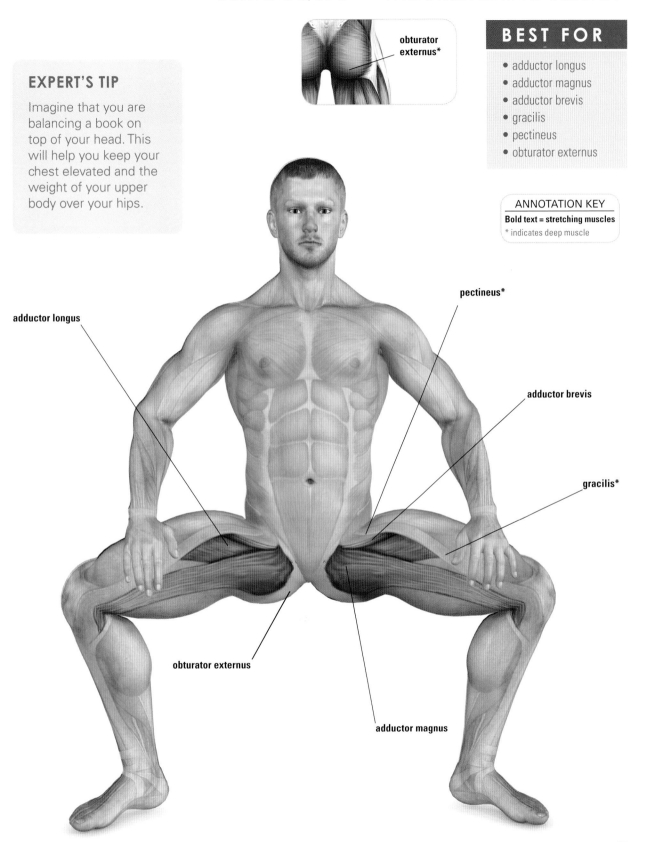

obturator
externus*

BEST FOR

- adductor longus
- adductor magnus
- adductor brevis
- gracilis
- pectineus
- obturator externus

ANNOTATION KEY
Bold text = stretching muscles
* indicates deep muscle

EXPERT'S TIP

Imagine that you are balancing a book on top of your head. This will help you keep your chest elevated and the weight of your upper body over your hips.

adductor longus

pectineus*

adductor brevis

gracilis*

obturator externus

adductor magnus

SIDE-LEANING SUMO SQUAT

1 Begin at the bottom position of the Sumo Squat (see pages 88–89), with your knees bent and thighs parallel to the floor.

2 Drop your right forearm onto your right thigh, just above the kneecap. Bring your left arm and hand straight up and reach over to the right side. Hold this position as you stretch.

3 Round your left arm down so that both forearms are on your thighs, and bring your head back to the center.

4 Push off with your heels at the bottom of the squat, squeezing your glutes and inner thighs to rise back to the starting position.

5 Switch sides, and repeat.

TARGETS
- Groin muscles
- Hip adductors

DO
- Keep your upper body and back straight.

AVOID
- Leaning forward as you stretch to the side.
- Letting your knees extend beyond your toes.
- Tensing your jaw, as this will restrict your breathing.

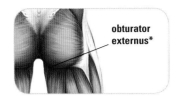

obturator externus*

BEST FOR

- adductor longus
- adductor magnus
- adductor brevis
- gracilis
- pectineus
- obturator externus

EXPERT'S TIP

Imagine that you are a puppet with a string attached to your head, lifting you up before you reach to the side. This will greatly increase the intensity of the stretch.

pectineus*

adductor brevis

obliquus externus

gracilis*

obliquus internus*

obturator externus

adductor magnus

adductor longus

ANNOTATION KEY

Bold text = stretching muscles
* indicates deep muscle

SIDE-LUNGE STRETCH

1 From the Sumo Squat (see pages 88–89), drop your hands onto the floor in front of you, transferring some of your weight onto your arms.

EXPERT'S TIP

Take the appropriate weight onto your hands to decrease the intensity of this stretch.

TARGETS
- Hip adductors
- Hip flexors
- Hamstrings
- Inner thighs
- Gluteal area

DO
- Aim to drop your gluteals toward the floor, as this will increase the intensity of the stretch.
- Keep the extended leg turned out from the hips, foot flexed.

AVOID
- Overdoing the stretch by extending too far to the side.

2 Slowly shift your body over to the right, staying as low as possible, bending your right knee, and extending and straightening your left leg.

3 Carefully return to the center, switch sides, and repeat.

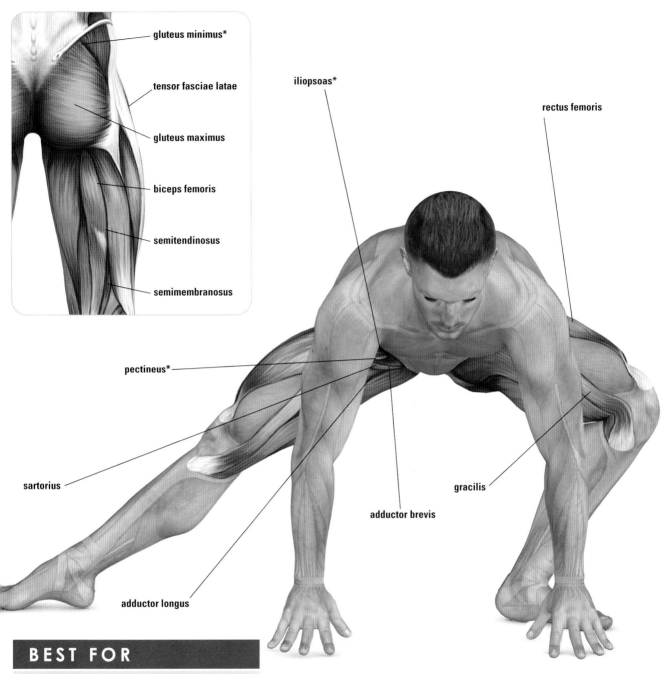

gluteus minimus*

tensor fasciae latae

gluteus maximus

biceps femoris

semitendinosus

semimembranosus

iliopsoas*

rectus femoris

pectineus*

sartorius

gracilis

adductor brevis

adductor longus

BEST FOR

- gluteus minimus
- tensor fasciae latae
- gluteus maximus
- iliopsoas
- rectus femoris
- sartorius
- pectineus
- adductor longus
- adductor brevis
- gracilis
- biceps femoris
- semitendinosus
- semimembranosus

ANNOTATION KEY

Bold text = stretching muscles

* indicates deep muscle

93

FORWARD LUNGE

❶ Begin in the Sumo Squat (see pages 88–89).

BEST FOR

- rectus femoris
- vastus lateralis
- vastus intermedius
- vastus medialis
- biceps femoris
- semitendinosus
- semimembranosus
- gluteus maximus
- adductor longus
- adductor magnus
- adductor brevis
- gracilis
- pectineus
- obturator externus
- iliopsoas
- gluteus minimus
- tensor fasciae latae

TARGETS
- Quadriceps
- Gluteal area
- Inner thighs
- Hamstrings
- Ball of the foot

DO
- Keep your back leg extended in line with your hips to form one long straight line.
- Keep your knee directly above your ankle.

AVOID
- Dropping your back extended leg to the floor.
- Hunching your shoulders.

❷ Drop your hands onto the floor in front of you, transferring some of your weight onto your arms.

❸ Carefully "walk" your hands to the right as you pivot your right foot forward.

Modification
Advanced: Place your palms or fingertips on the floor on either side of your front foot. Keep your head in line with your spine, focusing your gaze forward a few feet in front of you.

4 Step your left leg back behind your body, extending it straight, keeping your right knee bent.

5 Place your palms on your knee, and hold.

6 Return to the Sumo Squat, and repeat on the other side.

ANNOTATION KEY
Bold text = stretching muscles
* indicates deep muscle

iliopsoas*

pectineus

adductor brevis

adductor longus

vastus medialis

gracilis*

adductor magnus

tensor fasciae latae

vastus intermedius*

rectus femoris

gluteus minimus*

gluteus maximus

obturator externus

semitendinosus

biceps femoris

semimembranosus

vastus lateralis

FORWARD LUNGE WITH TWIST

1 From the Forward Lunge (see pages 94–95) with your right leg forward, place your hands on the floor on either side of your right foot.

BEST FOR

- obliquus externus
- obliquus internus
- rectus femoris
- vastus lateralis
- vastus intermedius
- vastus medialis
- biceps femoris
- semitendinosus
- semimembranosus
- gluteus minimus
- gluteus maximus
- adductor longus
- adductor magnus
- adductor brevis
- gracilis
- pectineus
- obturator externus
- iliopsoas
- tensor fasciae latae

2 Balance your weight on your left hand, and carefully and slowly guide your right arm up toward the ceiling, twisting your torso.

3 Return to the center, and repeat on the other side.

TARGETS
- Quadriceps
- Gluteal area
- Hip adductors
- Hamstrings
- Obliques
- Rib cage
- Chest
- Shoulders

DO
- Keep your focus up toward the elevated arm and hand, and point the fingers of the hand in the air.
- Keep your chest slightly elevated.
- Keep your legs and feet parallel.

AVOID
- Holding your breath.
- Rounding your back.

EXPERT'S TIP

Imagine that you are a puppet and a taut string is coming out from the top of your head, another one from the tip of your elevated middle finger, and another one from your extended leg's heel. Visualizing this image will elongate your whole body.

ANNOTATION KEY
Bold text = stretching muscles
* indicates deep muscle

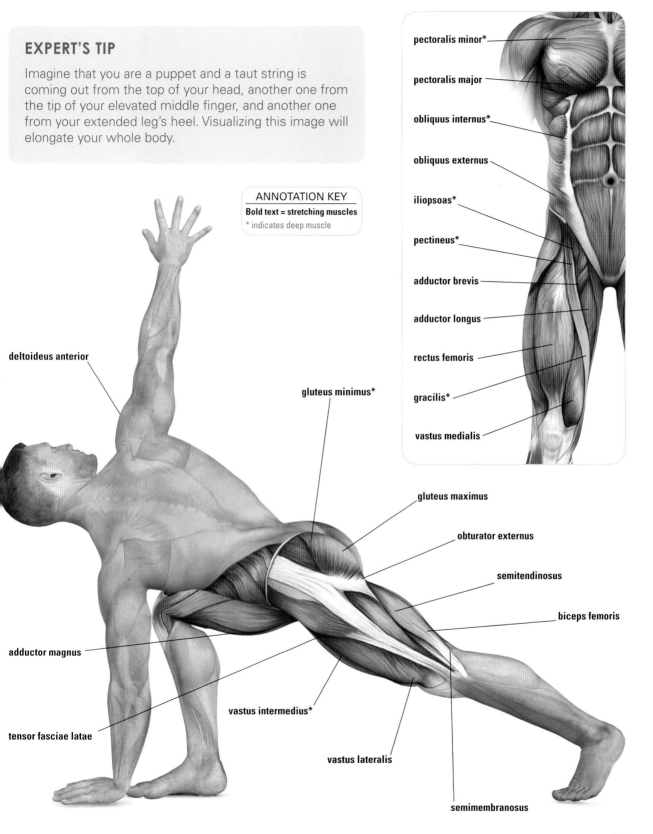

pectoralis minor*

pectoralis major

obliquus internus*

obliquus externus

iliopsoas*

pectineus*

adductor brevis

adductor longus

rectus femoris

gracilis*

vastus medialis

deltoideus anterior

gluteus minimus*

gluteus maximus

obturator externus

semitendinosus

biceps femoris

adductor magnus

vastus intermedius*

tensor fasciae latae

vastus lateralis

semimembranosus

STRAIGHT-LEG LUNGE

1 Stand with your legs and feet parallel and shoulder-width apart. Bend your knees very slightly and tuck your pelvis slightly forward, lift your chest, and press your shoulders downward and back.

EXPERT'S TIP

Keep your chest elevated, and focus your gaze toward your front foot; this will help elongate your torso and intensify the stretch in both the lower back and the hamstrings.

2 Take one step forward with the right foot.

3 Keeping your legs straight, lean your torso forward as far as possible over your right leg. Allow the weight of the upper body to intensify the stretch.

4 Return to standing, and repeat on the other side.

TARGETS
- Hamstrings
- Lower back
- Calves

DO
- Flex the front foot by lifting the ball of the foot off the floor to maximize the intensity of the stretch.
- Keep the heel of your back leg on the floor throughout the stretch.

AVOID
- Holding unnecessary tension in the upper body—relax and breathe in and out naturally.

BEST FOR

- biceps femoris
- semitendinosus
- semimembranosus
- erector spinae
- gastrocnemius
- soleus

Modification
Advanced: Place your hands flat on the floor on either side of the front foot.

ANNOTATION KEY
Bold text = stretching muscles
* indicates deep muscle

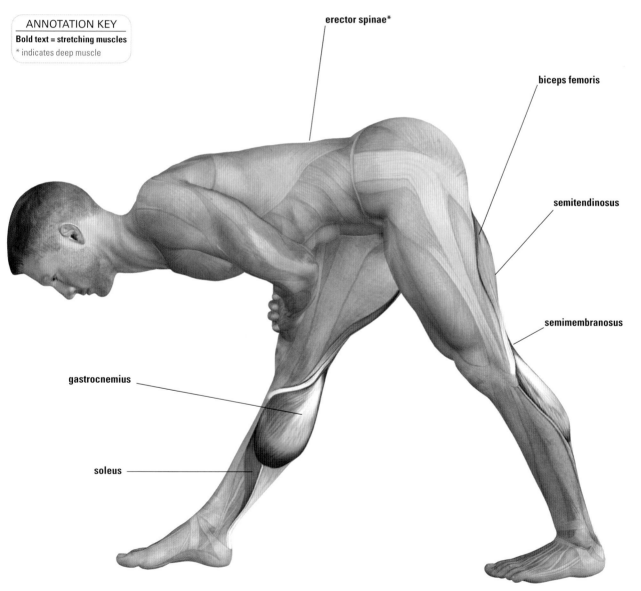

erector spinae*

biceps femoris

semitendinosus

semimembranosus

gastrocnemius

soleus

DOWNWARD-FACING DOG

1 Kneel on all fours, with your knees directly below your hips. Position your hands on the floor just beyond shoulder-width apart, slightly in front of your body, palms downward and fingertips facing forward.

2 Press against the floor, keeping your elbows straight. Lift your tailbone up toward the ceiling and your knees away from the floor. Lengthen your hips away from your rib cage to elongate your spine.

3 Press your heels toward the floor, and contract your thighs as you straighten your legs to form a V shape with your body. Broaden your chest and shoulders, and position your head between your arms.

TARGETS
- Hamstrings
- Calves
- Back
- Upper arms
- Chest
- Achilles tendon
- Gluteal region

DO
- Engage your entire hand fully into the floor at all times to avoid excess strain on your wrist joint.
- Keep your head in line with your spine.
- Keep your back flat and your chest elevated.

AVOID
- Holding your breath: relax your jaw slightly and breathe normally.

BEST FOR

- pectoralis major
- pectoralis minor
- serratus anterior
- triceps brachii
- deltoideus posterior
- intercostales interni
- intercostales externi
- biceps femoris
- semitendinosus
- semimembranosus
- erector spinae
- gastrocnemius
- soleus
- gluteus maximus

EXPERT'S TIP

The Downward-Facing Dog is one of the most widely known yoga poses. This basic pose has been known to calm the brain and helps relieve stress and mild depression.

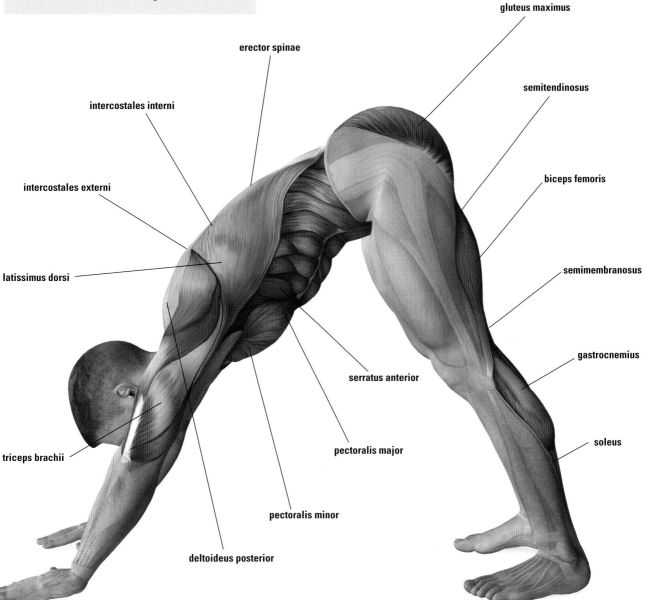

gluteus maximus

erector spinae

intercostales interni

semitendinosus

intercostales externi

biceps femoris

latissimus dorsi

semimembranosus

gastrocnemius

serratus anterior

triceps brachii

soleus

pectoralis major

pectoralis minor

deltoideus posterior

WIDE-LEGGED FORWARD BEND

1 Stand with your legs and feet parallel and generously outside of shoulder width. Bend your knees very slightly and tuck your pelvis slightly forward, lift your chest, and press your shoulders downward and back.

BEST FOR

- gluteus maximus
- gluteus medius
- gluteus minimus
- rectus abdominis
- transversus abdominis
- obliquus externus
- obliquus internus
- biceps femoris
- semitendinosus
- semimembranosus
- erector spinae
- gastrocnemius
- soleus

2 Bend forward from the hips, keeping your back flat.

TARGETS
- Hamstrings
- Lower back
- Gluteal area
- Calves

DO
- Keep your chest elevated.
- Exhale as you bend forward from the hips.

AVOID
- Tensing your neck and shoulders.

3 Place your fingertips or palms on the floor.

Modification

Easier: Widen your stance or place a yoga block on the floor for support.

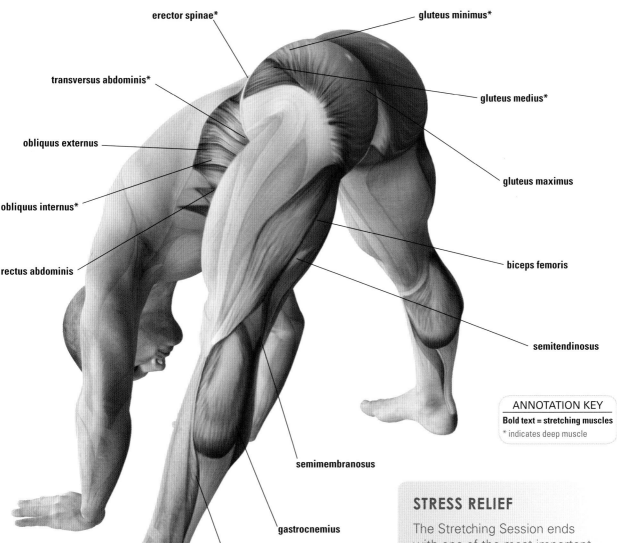

erector spinae*

gluteus minimus*

transversus abdominis*

gluteus medius*

obliquus externus

gluteus maximus

obliquus internus*

rectus abdominis

biceps femoris

semitendinosus

semimembranosus

gastrocnemius

soleus

ANNOTATION KEY
Bold text = stretching muscles
* indicates deep muscle

STRESS RELIEF

The Stretching Session ends with one of the most important stretches: the Wide-Legged Forward Bend. Not only is it the most effective way to lengthen the hamstrings and back muscles, it also allows for a more fluid range of movement.

Dancers also rely on this stretch to relieve the stress of those last seconds standing in the wings, ready to go onstage, or before an audition. You can perform this stretch when you are stressed, when you need to calm down, or when you just want to take a short break in your day.

Modification

Advanced: "Walk" your hands in between your legs, bend your elbows, and gently place your forehead on the floor. Your hands should be available if necessary to balance.

PARTNER STRETCHES

To stay on track with a fitness program, accountability—that state of being responsible to someone or for some action—is a necessity.

Respect yourself as you would a business client or any other outside obligation, and stay true to your schedule of workouts and stretching. By including a partner or friend in your fitness program, you will further ensure that you keep to that time slot of self-care.

There are so many benefits of stretching with a partner—this list features just a few of them.

- Working with a partner helps you achieve a greater degree of flexibility and range of motion.
- The desire to impress your partner can prove motivating—and this motivation challenges you to work through personal plateaus and struggles.
- You're less likely to skip the partner stretch session.
- Stretching with a partner is a perfect opportunity to combine social time with fitness. It lets you condense your day to fit everything into your schedule!
- A partner can serve as an extra set of eyes, correcting form and giving you a critique of your performance.
- A partner can help relieve the monotony of a repetitious program, making it fresh and exciting.
- Working with a partner empowers you to spread the information and knowledge that's in this book to friends and loved ones. Your self-confidence increases when you share what you know!

CHOOSE YOUR PARTNER

If you are at all apprehensive about partner stretching, it's a good idea to start by working with a trained professional, such as a personal trainer at your local gym or a physical therapist. Once you feel secure with the techniques, try stretching with friends or family members.

Any physical activity program is a special thing to share with someone. Choose a partner, loved one, friend, coworker, relative—anyone with whom you feel comfortable spending quality time.

How to Stretch with a Partner

Feel free to execute any of the stretches in this book with a stretching partner. Some of the following stretches focus on a specific technique called PNF (proprioceptive neuromuscular facilitation), which combines passive stretching with isometric contractions. You can perform the PNF technique on your own, but a partner greatly helps. There are a number of different methods, but for the purposes of this book we are going to focus on the hold-relax method.

Hold-Relax

To understand the hold-relax techniques, take a look at how you would perform a hamstring stretch with a partner.

- The partner moves the stretcher's extended leg to the point of a comfortable yet challenging stretch. This passive stretch is held for 10 seconds.

- The stretcher then isometrically contracts the hamstrings by pushing the extended leg against the partner's hand. The partner provides just enough force so that the leg remains static. This is the "hold" phase, which will be held for 6 seconds.

- The stretcher will then "relax," and the partner completes a second passive stretch, this time held for 30 seconds. For the second stretch, the stretcher's extended leg should have a greater range of motion than it had while performing the first passive stretch.

WORKING TOGETHER

The partner will provide proper leverage and control over the stretch, keeping safety as the priority. As a precaution, this book focuses only on larger muscles, such as the legs and trunk of the body.

Remember: Stretching should *never* hurt.

ASSISTED BUTTERFLY STRETCH

1 STRETCHER: Sit on the floor, bringing your heels together. Keep them a comfortable distance from your core.

2 STRETCHER: Pull your upper body forward to where you will feel a stretch in your groin, as well as the inside of your upper thighs, and then reach forward, placing the palms of your hands on the floor in front of you.

3 PARTNER: Stand behind the stretcher, and place your hands on the stretcher's inner thighs, close to the knees. Gently add pressure, and hold for 10 seconds.

TARGETS
• Hip adductor
• Inner thighs

DO
• As partner, proceed with caution, go slow, and communicate with the stretcher.

AVOID
• As partner, avoid pressing downward with your knees as you lean forward to apply pressure to the stretcher's inner thighs.

4 Both relax, and then repeat, holding for 30 seconds.

5 Again relax, and repeat for a second set.

6 Switch roles, and perform the entire sequence for two sets.

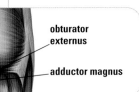

pectineus*

adductor longus

adductor brevis

gracilis*

obturator externus

adductor magnus

BEST FOR
• adductor longus
• adductor magnus
• adductor brevis
• gracilis
• pectineus
• obturator externus

ANNOTATION KEY
Bold text = stretching muscles
* indicates deep muscle

ASSISTED HAPPY BABY

1 STRETCHER: Lie on your back, with both legs elongated and parallel and your arms extended away from your torso, palms facing up.

2 PARTNER: Lift one leg over the stretcher's body, and position your feet next to either side of the stretcher's upper rib cage.

BEST FOR

- gluteus maximus
- gluteus medius
- gluteus minimus
- piriformis
- biceps femoris
- semitendinosus
- semimembranosus
- erector spinae
- multifidus spinae

3 STRETCHER: Slightly bend your legs at the knees, and lift up one leg at a time so that your partner can grasp each ankle. Turn out your legs from the hips.

4 PARTNER: Bring the stretcher's ankles to the front of your body, wrapping them around your thighs. Hold for 20 to 30 seconds.

5 Relax, and repeat for a second set.

6 Switch roles, and perform the entire sequence for two sets.

TARGETS
- Hips
- Inner thighs
- Lower back

DO
- As partner, carefully watch where you are stepping and placing your feet.
- As stretcher, flex your feet.

AVOID
- As stretcher, avoid holding our breath.

Modification
Advanced: Place your palms on the soles of the stretcher's flexed feet, and gently press downward.

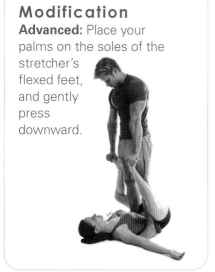

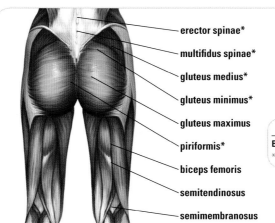

erector spinae*

multifidus spinae*

gluteus medius*

gluteus minimus*

gluteus maximus

piriformis*

biceps femoris

semitendinosus

semimembranosus

ANNOTATION KEY
Bold text = stretching muscles
* indicates deep muscle

ASSISTED UNILATERAL THIGH STRETCH

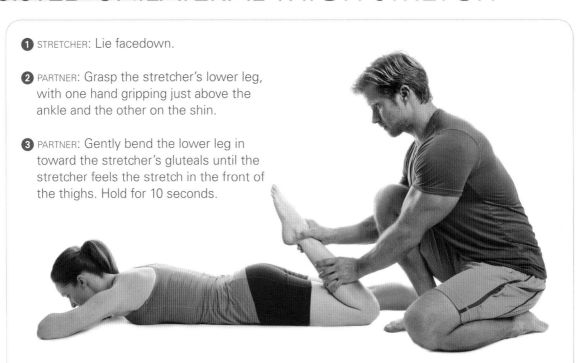

① STRETCHER: Lie facedown.

② PARTNER: Grasp the stretcher's lower leg, with one hand gripping just above the ankle and the other on the shin.

③ PARTNER: Gently bend the lower leg in toward the stretcher's gluteals until the stretcher feels the stretch in the front of the thighs. Hold for 10 seconds.

TARGETS
- Hip flexors
- Groin muscles

DO
- As partner, apply gentle pressure, never forcing the stretcher's foot down farther than is comfortable.

AVOID
- As stretcher, avoid holding your breath.

④ STRETCHER: While your partner provides appropriate resistance, try to push your leg back for 6 seconds, and then relax.

⑤ PARTNER: Again, gently bend the lower leg in toward the stretcher's gluteals until the stretcher feels the stretch in the front of the thighs. Hold for 30 seconds.

⑥ Again relax, and repeat for a second set.

7 Switch sides, and repeat for two sets.

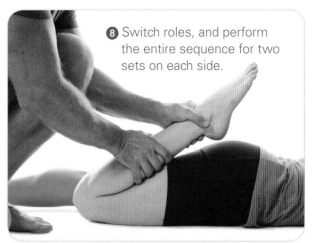

8 Switch roles, and perform the entire sequence for two sets on each side.

iliopsoas*

iliacus*

pectineus*

sartorius

BEST FOR

- rectus femoris
- iliopsoas
- iliacus
- sartorius
- tensor fasciae latae
- pectineus

ANNOTATION KEY

Bold text = stretching muscles
* indicates deep muscle

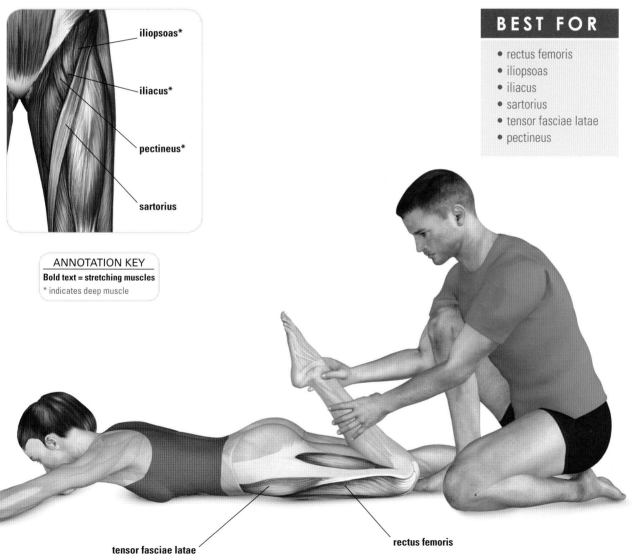

tensor fasciae latae

rectus femoris

ASSISTED UNILATERAL LEG RAISE

1 STRETCHER: Lie on your back, with your partner standing to your right, facing you. Lift your right leg toward the ceiling, so that your partner can take hold of it.

2 PARTNER: Take hold of your partner's lifted leg, and position yourself so that the lower calf rests comfortably on your right shoulder. Place your other hand on the thigh just above the kneecap.

3 PARTNER: Gently lean in toward the stretcher, providing an appropriate stretch. Hold for 10 seconds.

TARGETS
- Hamstrings
- Calves
- Gluteal region

DO
- As stretcher, concentrate on keeping your elevated leg as straight as possible.

AVOID
- As stretcher, avoid lifting your gluteals off the floor.
- As stretcher, avoid holding your breath.

4 STRETCHER: While your partner provides appropriate resistance, push your leg into your partner's shoulders. Hold for 6 seconds, and then relax.

5 PARTNER: Again gently lean in toward the stretcher, providing an appropriate stretch. Hold for 30 seconds.

6 Again relax, and repeat for a second set.

7 Switch roles, and perform the entire sequence for two sets.

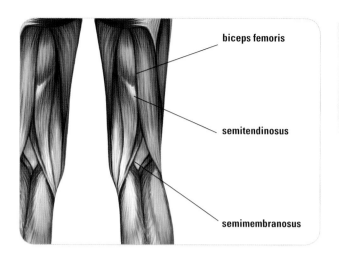

biceps femoris

semitendinosus

semimembranosus

EXPERT'S TIP

Make sure that you don't "sickle" your foot, which is when your ankle turns in so that your big toe points in toward the other foot.

BEST FOR

- biceps femoris
- semitendinosus
- semimembranosus
- gluteus maximus
- gluteus medius
- gluteus minimus
- gastrocnemius
- soleus

ANNOTATION KEY

Bold text = stretching muscles
* indicates deep muscle

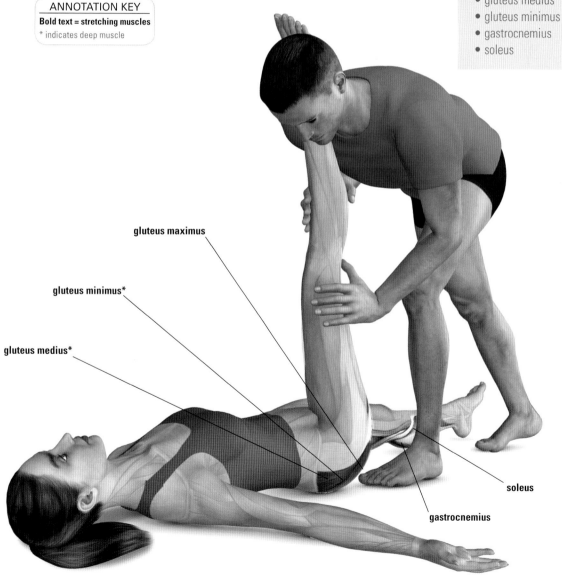

gluteus maximus

gluteus minimus*

gluteus medius*

soleus

gastrocnemius

ASSISTED CHEST STRETCH

1 STRETCHER: Sit on the floor, bringing your heels together. Keep them a comfortable distance from your core. Clasp your hands together behind your head.

2 PARTNER: Stand behind the stretcher, and bend and slightly knock your knees inward, placing them on the sides of the stretcher's middle back.

3 PARTNER: Reach up and over the stretcher's head, placing the inside of your forearms on the inside of the stretcher's upper forearms and inside of the biceps.

TARGETS
- Chest
- Shoulders

DO
- As partner, place your knees on either side of the stretcher's spine, never directly on the spine.

AVOID
- As partner, avoid putting too much pressure on the stretcher's back with your knees— use your knees only to provide stability.

4 PARTNER: Pull the stretcher's arms in toward yourself, while still providing stability with your knees on the middle back. Hold for 10 seconds, and then relax.

5 STRETCHER: While your partner provides appropriate resistance, push your arms against this resistance. Hold for 6 seconds, and then relax.

6 PARTNER: Again, pull the stretcher's arms in toward yourself, while still providing stability with your knees on the middle back. Hold for 30 seconds.

7 Again relax, and repeat for a second set.

8 Switch roles, and perform the entire sequence for two sets.

BEST FOR

- pectoralis major
- pectoralis minor
- deltoideus anterior

COMMUNICATION

Communication is the key to any relationship, and the same applies with your stretching partner. The stretcher must communicate the feeling of the stretch—what feels appropriate and what does not. The partner must make sure to ask questions and check how the stretcher is feeling.

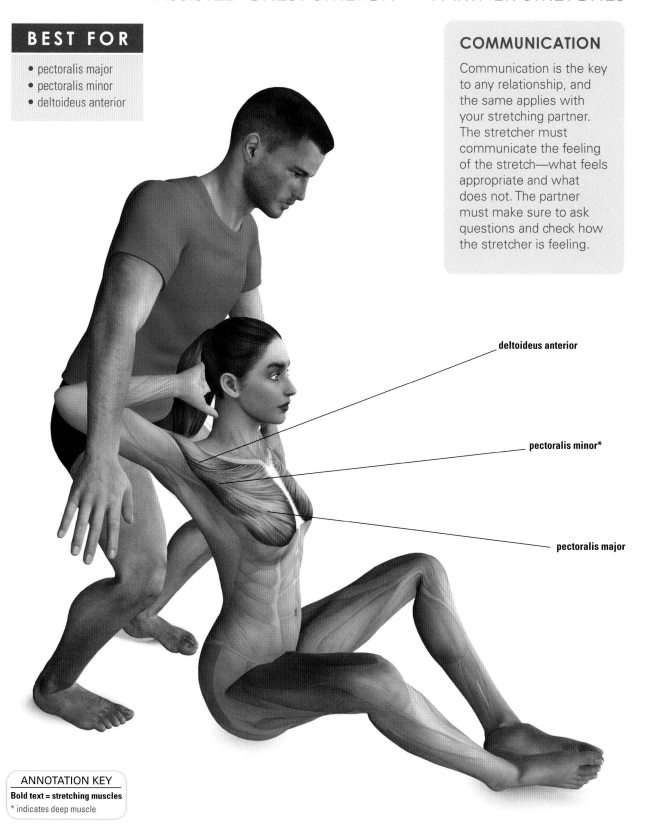

deltoideus anterior

pectoralis minor*

pectoralis major

ANNOTATION KEY
Bold text = stretching muscles
* indicates deep muscle

ASSISTED SEATED FORWARD BEND

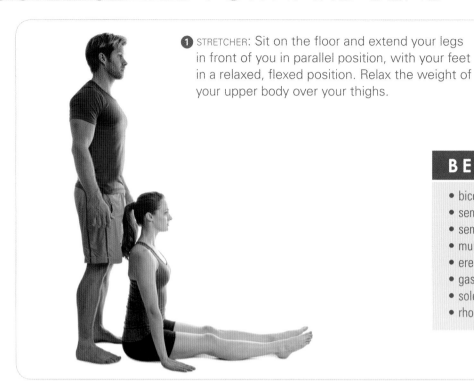

1 STRETCHER: Sit on the floor and extend your legs in front of you in parallel position, with your feet in a relaxed, flexed position. Relax the weight of your upper body over your thighs.

BEST FOR

- biceps femoris
- semitendinosus
- semimembranosus
- multifidus spinae
- erector spinae
- gastrocnemius
- soleus
- rhomboideus

TARGETS
- Hamstrings
- Lower back
- Upper back
- Calves

DO
- As stretcher, you can achieve a deeper stretch by flexing your feet and, if possible, reaching your hands around the bottom of your heels.
- As stretcher, keep your forearms lying above your kneecaps.

AVOID
- As partner, avoid bouncing downward—all of your movements should be steady and gentle.

2 PARTNER: Stand behind the stretcher and bend your legs so that your shins lightly rest on the stretcher's lower back. Place the palms of your hands on the stretcher's shoulder blades.

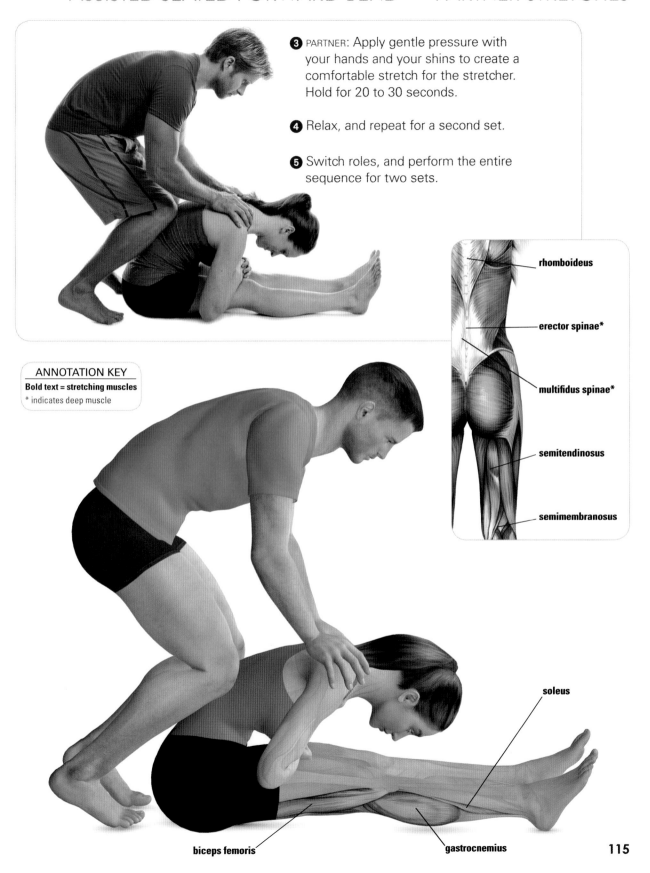

3 PARTNER: Apply gentle pressure with your hands and your shins to create a comfortable stretch for the stretcher. Hold for 20 to 30 seconds.

4 Relax, and repeat for a second set.

5 Switch roles, and perform the entire sequence for two sets.

ANNOTATION KEY
Bold text = stretching muscles
* indicates deep muscle

rhomboideus

erector spinae*

multifidus spinae*

semitendinosus

semimembranosus

soleus

biceps femoris

gastrocnemius

ASSISTED CHILD'S POSE

1 STRETCHER: Kneel with your buttocks resting lightly on your heels.

2 STRETCHER: Move your knees slightly out to the sides, and reach your arms forward to rest your palms flat on the floor.

3 PARTNER: Come to stand in front of the stretcher, and then bring your feet to either side of the stretcher's shoulders. Reach down and place your palms on the stretcher's outer thighs.

4 PARTNER: Gently and slowly apply a slight downward and backward pressure to the stretcher's lower body. Hold for 20 to 30 seconds.

5 Relax, and repeat for a second set.

6 Switch roles, and perform the entire sequence for two sets.

TARGETS
- Hip adductors
- Inner thighs
- Lower back
- Middle back

DO
- As stretcher, rest your forehead on a towel or mat.

AVOID
- As partner, avoid applying additional pressure, unless you have the stretcher's approval.

pectineus*
adductor longus
gracilis

latissimus dorsi
erector spinae*
multifidus spinae*
obturator externus
adductor magnus

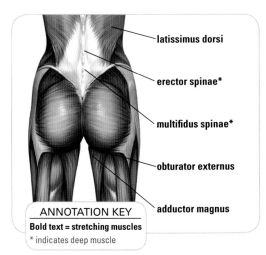

BEST FOR
- adductor longus
- adductor magnus
- gracilis
- pectineus
- obturator externus
- multifidus spinae
- erector spinae
- latissimus dorsi

ANNOTATION KEY
Bold text = stretching muscles
* indicates deep muscle

ASSISTED PRETZEL STRETCH

1 STRETCHER: Lie on your back, with both legs elongated and parallel and your arms extended away from your torso, palms facing up.

2 STRETCHER: Bend your right leg, placing the sole of your foot on the floor.

3 PARTNER: Kneel to the stretcher's right.

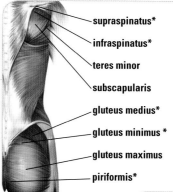

- supraspinatus*
- infraspinatus*
- teres minor
- subscapularis
- gluteus medius*
- gluteus minimus *
- gluteus maximus
- piriformis*

ANNOTATION KEY
Bold text = stretching muscles
* indicates deep muscle

- pectoralis minor*
- pectoralis major

4 STRETCHER: Carefully lift your buttocks off the floor, tilting your torso 2 to 3 inches to your left.

5 STRETCHER: Cross and bend your right leg to the left.

BEST FOR

- supraspinatus
- infraspinatus
- teres minor
- subscapularis
- gluteus maximus
- gluteus medius
- gluteus minimus
- piriformis
- pectoralis major
- pectoralis minor

6 PARTNER: Place your left hand on top of the stretcher's right shoulder, and your right hand on the stretcher's knee. With your left hand, gently press the stretcher's shoulder downward, while with your left hand press downward on the stretcher's right knee. Hold for 30 seconds.

7 Relax, and repeat for a second set.

8 Switch roles, and perform the entire sequence for two sets.

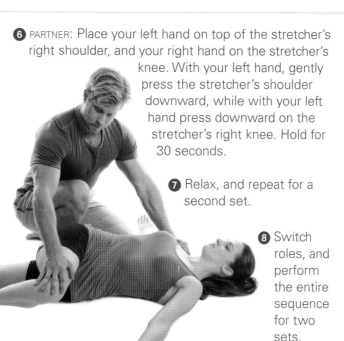

TARGETS
- Shoulders
- Gluteal region
- Chest

DO
- As stretcher, try to keep both shoulder blades in contact with the floor.
- As stretcher, keep your elbows and wrists lower than your shoulder—this will protect your rotator cuff.

AVOID
- As partner, avoid applying unnecessary pressure—you want only to increase the stretcher's desired personal stretch.

RUSSIAN SPLIT SWITCH

1 STRETCHER: Sit upright with your legs spread as widely as is comfortable, with your feet slightly flexed and your legs turned out from the hips so that your toes point upward.

2 STRETCHER: Find and place your hip bones on the floor.

3 PARTNER: Sit on the floor facing the stretcher, and extend your legs out to the sides so that the soles of your feet rest above the stretcher's inner ankles.

EXPERT'S TIP

When you are the stretcher, focus on utilizing your partner's feet placement above your ankles to aid in the turnout from your hips.

TARGETS
• Hamstrings
• Hip adductors

DO
• As stretcher, keep your back as flat as possible, elevate your chest, and maintain a long neck in line with the entire spine.

AVOID
• As either stretcher or partner, avoid holding your breath.

4 PARTNER: Reach out and take the stretcher's hands in yours.

5 PARTNER: Lean back slightly, taking the stretcher with you. Hold for 20 to 30 seconds.

6 Relax, and repeat for a second set.

7 Switch roles, and perform the entire sequence for two sets.

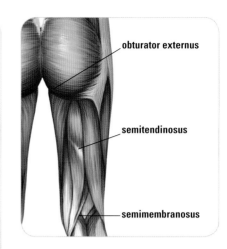

obturator externus

semitendinosus

semimembranosus

BEST FOR

- biceps femoris
- semitendinosus
- semimembranosus
- adductor longus
- adductor magnus
- gracilis
- pectineus
- obturator externus

ANNOTATION KEY
Bold text = stretching muscles
* indicates deep muscle

pectineus*

adductor longus

gracilis*

biceps femoris

adductor magnus

PREGNANCY STRETCHES

Should you begin or continue a stretching program if you are pregnant? The answer is a resounding yes.

To prepare for childbirth, a woman's body undergoes a host of changes, including the shifting of your center of gravity, realigning your posture, and loosening some of your joints, ligaments, and muscles. Many of the changes pregnancy makes on the body can result in aches and pains and a general feeling of immobility or estrangement from your own body. Stretching helps you stay in touch with your body, and it can help relieve some of those aches and pains, keeping you limber and lithe.

Your Changing Body
Among the most noticeable changes is your center of gravity shift. To compensate, the muscles in your chest, lower back, and hips tighten. Stretches that focus on posture and balance are an excellent way to combat this tightness.

During pregnancy, levels of the hormone relaxin rise in a woman's body. This hormone, which is believed to soften the pubic symphysis in the pelvis and facilitate labor, may be produced up to three months after childbirth. It relaxes ligaments, as well as muscles, and makes pregnant women and new mothers vulnerable to overstretching. Relaxin levels may also remain high after a miscarriage, which can put you at risk for stretching beyond your normal and healthy range.

Stretching During Pregnancy
For most healthy women experiencing normal pregnancies, stretching has numerous benefits, including:

- It relaxes the body and prepares it for delivery.
- It allows you to practice breathing.
- It helps to ease stress.

After Pregnancy
After your baby is born, your body will again undergo changes. Adhering to a regular stretching program can help you adjust. Stretching also helps ease the aches or stiffness that often accompany caring for an infant. For instance, holding your newborn and breast-feeding can make your neck extremely stiff. Regularly perform the Neck Stretches (see pages 76–77) to keep your neck feeling loose.

While allowing you time to focus on your own needs, post-pregnancy stretching rebalances your muscles, helps you to avoid injury, reduces stress, and rebuilds your body image.

The "Baby Bounce-Back"

You can achieve the body you had prior to pregnancy—and even improve it—by following a few simple dietary guidelines.

- Eat a nutrient-rich diet. Eating healthy isn't always easy with a baby, so plan ahead and have nutritious snacks readily available.
- Eat high-protein foods. The protein will help you build necessary muscle to make your workouts more intense and powerful. This, in turn, speeds up your metabolism, which aids in weight loss.
- Eat fiber-rich foods. Foods high in fiber help make you feel full sooner. Fiber also helps fat pass through the digestive system.

STRETCHING SAFELY BEFORE AND AFTER CHILDBIRTH

Always consult and be supervised by your medical specialist prior to engaging in any type of workout program. Keep in mind that even as your changing body puts you in the position to derive enormous benefits from stretching, you must also proceed carefully.

- Never bounce with these stretches; this can cause serious injury. Do not overstretch.
- During your third trimester, consult with your doctor before attempting and performing any stretches that require you to lie on your back. This may cause shortness of breath and dizziness.
- For post-pregnancy stretches, you should concentrate on those that use a "turned in" position of the legs.
- Avoid sitting cross-legged for up to three months after giving birth, and avoid butterfly positioning of the legs. When possible, keep your knees together.

To really achieve the Baby Bounce-Back, begin a full workout program that includes stretching exercises, along with toning exercises and cardiovascular exercises. Don't just concentrate on crunches—try to focus more on high-intensity activities, such as cycling, jumping rope, swimming, and running.

TORSO ROTATION

① Sit on the floor with your legs extended in front of you and turned out slightly beyond shoulder width.

② Rest your hands on the floor behind you, and lean back slightly.

③ Slowly raise your left arm upward, just in front of your head, with your elbow slightly bent and your palm facing inward.

④ Turn your head, gazing to the right as you move your left arm over to right, aiming your hand slightly behind your body, creating a gentle stretch in your ribs and back.

TARGETS
- Middle back
- Lower back
- Obliques

DO
- When stretching toward the right, reach toward the back right corner behind you—and vice versa.
- Keep your supporting elbow slightly bent.
- Keep your feet flat on the floor, your legs in parallel, and your chest lifted.

AVOID
- Lifting your shoulders up toward your ears—you want to keep your neck elongated.

5 Return to the starting position, and repeat on the other side.

EXPERT'S TIP

Think about the position of your raised arm. Your arm should be kept slightly in front of your head, with your elbow slightly bent so that you feel energy extending out of your fingers.

rhomboideus*

erector spinae*

latissimus dorsi

obliquus externus

obliquus internus*

ANNOTATION KEY

Bold text = stretching muscles

* indicates deep muscle

HAND-ON-KNEE STRETCH

1 Sit on the floor with your legs extended in front of you, your feet relaxed and flexed slightly. Bend your right leg, and rest the sole of your foot on your left inner thigh.

2 Place the palms of your hands on top of your left thigh, just above your kneecap.

3 Gently lean over your left leg until you feel a comfortable stretch in your hamstring.

4 Return to the starting position, and repeat on the other side.

EXPERT'S TIP
Maintain length in your neck, with your chin lifted slightly higher than you would naturally hold it.

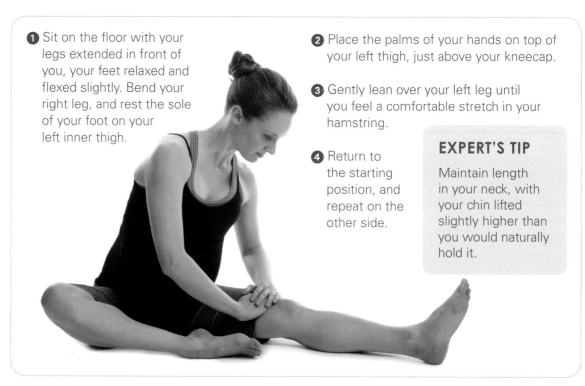

TARGETS
- Lower back
- Hamstrings
- Calves

DO
- Keep your chest lifted.
- Aim to decrease the space under the kneecap of your extended leg.
- Keep your shoulders pressed slightly and gently downward, away from your ears. Place one hand on your lower back to guard against strain, if necessary.

AVOID
- Lifting and holding tension in your bent knee.

BEST FOR
- biceps femoris
- semitendinosus
- semimembranosus
- erector spinae
- multifidus spinae
- gastrocnemius
- soleus

ANNOTATION KEY
Bold text = stretching muscles
* indicates deep muscle

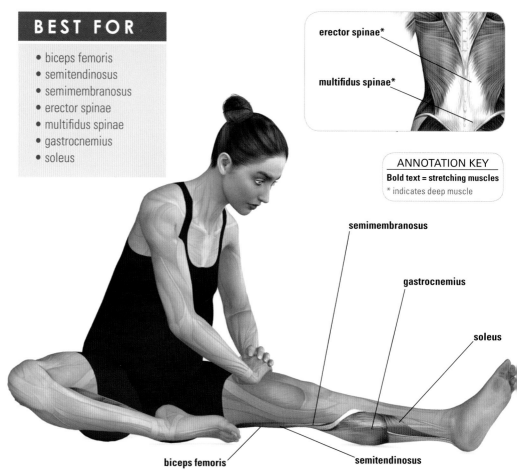

erector spinae*

multifidus spinae*

semimembranosus

gastrocnemius

soleus

biceps femoris

semitendinosus

LYING PELVIC TILT

❶ Lie on your back with your knees bent. Your feet should be flat on the floor, and your legs in parallel.

❷ Place your hands comfortably on your belly.

❸ Slightly and carefully arch your lower back.

❹ Rotate your pelvis forward, which will flatten your lower back onto the floor.

❺ Return to the starting position, and repeat if desired.

TARGETS
• Lower back

DO
• Keep your chest slightly elevated.
• Relax your jaw.
• Breathe normally throughout the exercise.

AVOID
• Performing this stretch during your third trimester. During your first and second trimesters, proceed with caution and stop immediately if you feel any discomfort.

BEST FOR

• erector spinae
• multifidus spinae

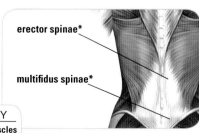

erector spinae*

multifidus spinae*

ANNOTATION KEY
Bold text = stretching muscles
* indicates deep muscle

UNILATERAL GOOD MORNING STRETCH

① Stand with your legs and feet parallel and shoulder-width apart. Bend your knees very slightly. Your pelvis should be slightly tucked, your chest lifted, and your shoulders pressed gently downward away from your ears.

② Place your right hand on your upper thigh. With your left arm, reach up toward the ceiling, palm inward.

③ Carefully lean to the right.

④ Return to the starting position, and repeat on the other side.

TARGETS
- Neck
- Shoulders
- Rib cage

DO
- Keep your head in line with your spine by lifting your chin slightly.

AVOID
- Moving your lower body.

EXPERT'S TIP

Make sure that the distance between your head and raised arm stays the same, whether you are standing up straight or leaning over to stretch.

trapezius

deltoideus posterior

intercostales interni*

intercostales externi

BEST FOR

- trapezius
- intercostales externi
- intercostales interni
- deltoideus posterior

ANNOTATION KEY
Bold text = stretching muscles
* indicates deep muscle

CAT STRETCH

1 Kneel on all fours, with your hands at shoulder width and your knees spread 2 to 3 inches apart.

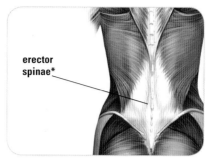

erector spinae*

ANNOTATION KEY

Bold text = stretching muscles

* indicates deep muscle

BEST FOR

- erector spinae

---

2 Round your spine upward as you draw your navel in toward your spine, keeping your hips lifted and your shoulders stable.

3 Hold the stretch at the top, and release.

TARGETS
- Back

DO
- Push down with your hands and knees to achieve maximum contraction.

AVOID
- Tensing your neck and shoulders.
- Hyperextending your lower back or arms.
- Holding your breath.

DOWNWARD-FACING DOG

1 Stand with your legs and feet parallel and shoulder-width apart. Bend your knees very slightly, and carefully fold forward to touch your fingertips to the floor.

2 Bend your knees slightly, and tuck your pelvis slightly forward, lift your chest, and press your shoulders downward and back.

3 Slowly "walk" your hands forward as you lift your tailbone up toward the ceiling.

TARGETS
- Hamstrings
- Calves
- Back
- Upper arms
- Chest
- Achilles tendon
- Gluteal region

DO
- Engage your entire hand fully into the floor at all times to avoid excess strain on your wrist joint.
- Keep your head in line with your spine.
- Keep your back flat and your chest elevated.

AVOID
- Holding your breath: relax your jaw slightly and breathe normally.

CLASSIC STRETCH

The Downward-Facing Dog offers relief from stiffness in the neck, legs, calves, shoulders, and lower back. This stretch, also a well-known yoga pose, helps expel carbon dioxide from the lungs, making room for oxygen to enter and rejuvenate the whole body.

The Downward-Facing Dog also decreases neck tension and headaches, often associated with pregnancy, by elongating the cervical spine and the neck.

4 Press your heels toward the floor, and contract your thighs as you straighten your legs to form a V shape with your body. Broaden your chest and shoulders, and position your head between your arms.

BEST FOR

- pectoralis major
- pectoralis minor
- serratus anterior
- triceps brachii
- deltoideus posterior
- intercostales interni
- intercostales externi
- biceps femoris
- semitendinosus
- semimembranosus
- erector spinae
- gastrocnemius
- soleus
- gluteus maximus

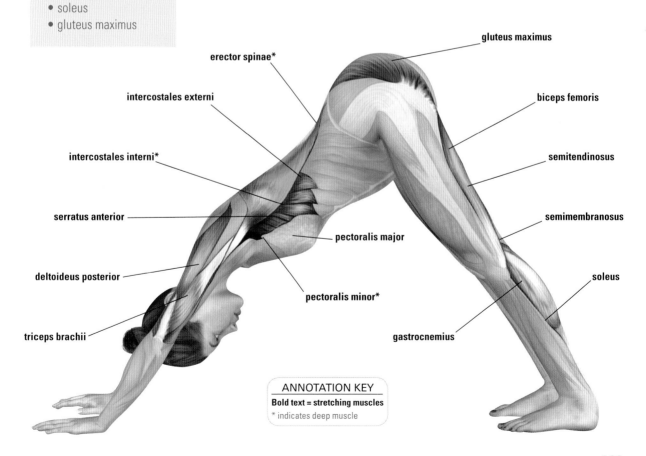

erector spinae*

intercostales externi

intercostales interni*

serratus anterior

deltoideus posterior

triceps brachii

gluteus maximus

biceps femoris

semitendinosus

semimembranosus

soleus

pectoralis major

pectoralis minor*

gastrocnemius

ANNOTATION KEY
Bold text = stretching muscles
* indicates deep muscle

OFFICE STRETCHES

So many of us spend much of our working lives sitting at a desk, looking up at a computer screen. Others spend all day on their feet or repeat the same movements over and over. Stretching can help alleviate the stiffness and discomfort that comes from performing these tasks.

Your Job and Your Health

If you take care of yourself, it shows that you are capable of taking care of business—and taking care of the company you work for. A fit, healthy body works more efficiently than one that is out of shape, and it is better at surviving and processing stress, not to mention coping with heavy workloads and other stressful situations.

Doing jumping jacks in the office may not be feasible, but almost all of us can find a few minutes a day to reenergize with the simple, inconspicuous stretches from an office chair demonstrated on the following pages.

You can perform any of these stretches individually or as a flow. Train yourself to take 30- to 60-second breaks every 30 minutes. Divide up your stretching session, getting in a few of these stretches throughout your day.

For those of you who can spare a bit more time during the workday, the Workplace Stretch Routine (shown opposite) will recharge your body and your mind.

OFFICE CHAIR OR FITNESS BALL?

Many companies now allow their employees to replace "regular" office chairs with fitness balls or specialized balance ball chairs. These work to combat the ill effects of a sedentary work environment. Rather than passively sitting in a stable seat, sitting on an unstable ball engages your core—your abdominals, obliques, and lower back must work to keep you positioned in the chair and for you to maintain correct posture.

WORKPLACE STRETCH ROUTINE

1 Hold for **10 seconds** each side

Side Neck Tilt
(page 76)

2 Hold for **20 seconds**

Back-of-the-Neck Stretch
(page 77)

3 Hold **5 seconds** for 2 sets

Lion Stretch
(page 75)

4 Hold **10 seconds** for 3 sets

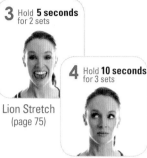

Eye Box Stretch
(page 75)

5 Hold for **10 seconds** each side

Forearm Stretches
(pages 82–83)

6 Hold for **20 seconds**

Good Morning Stretch
(pages 72–73)

7 Hold for **20 seconds** each side

Wall-Assisted Chest Stretch
(pages 80–81)

8 Hold for **20 seconds**

Standing Back Roll
(page 71)

9 Hold for **20 seconds** each side

Standing Quadriceps Stretch
(page 86)

10 Hold for **20 seconds** each side

Toe-Up Calf Stretch
(page 85)

11 Hold for **20 seconds**

Toe Touch
(page 70)

12 Hold for **20 seconds**

Sumo Squat
(pages 88–89)

13 Hold for **20 seconds** each side

Side-Leaning Sumo Squat
(pages 90–91)

SEATED TWISTS

1 Sit upright on a chair, with your legs separated and your feet planted firmly on the floor.

2 Keeping your back straight and your chest open, rotate your torso to the right.

3 Return to the starting position, and repeat on the other side.

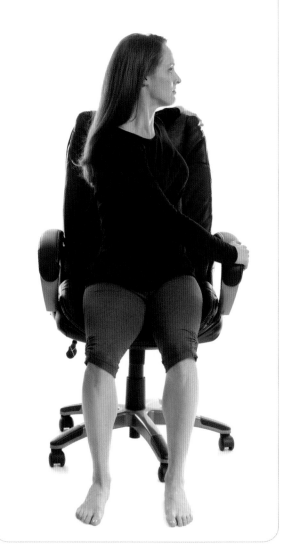

TARGETS
- Obliques
- Middle back

DO
- Twist from the hips, keeping your lower body stable.

AVOID
- Allowing your buttocks to rise from the chair as you twist to the side.

ANNOTATION KEY

Bold text = stretching muscles
* indicates deep muscle

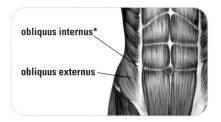

obliquus internus*

obliquus externus

latissimus dorsi

SEATED FIGURE 4

BEST FOR

- piriformis
- gluteus maximus
- gluteus medius
- gluteus minimus
- erector spinae
- quadratus femoris

1 Sit upright on a chair, with your legs separated and your feet planted firmly on the floor.

2 Place your right ankle on your left knee.

3 Lean forward from the hips until you feel a stretch in your buttocks and lower back.

4 Return to the starting position, and repeat with your left leg crossed over the right.

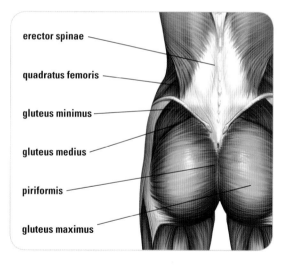

erector spinae

quadratus femoris

gluteus minimus

gluteus medius

piriformis

gluteus maximus

TARGETS
- Gluteal region

DO
- Bend only as far as is comfortable.

AVOID
- Allowing your buttocks to rise from the chair as you lean forward.

Modification

Advanced: Continue from step 3 to fold completely forward, touching your fingertips to the floor in front of you.

FORWARD BEND HIP SHIFT

1 Sit upright on a chair, with your legs widely separated and your feet planted firmly on the floor.

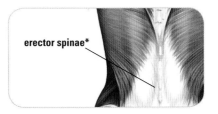

erector spinae*

2 Keeping your knees bent, bring your chest toward your thighs to rest your hands on floor.

3 Hold the downward position, and then return to the starting position, and repeat.

TARGETS
- Back
- Inner thighs

DO
- Keep your buttocks planted in the chair as you lean forward.

AVOID
- Dropping your head too quickly; keep your movements slow and controlled.

BEST FOR

- iliopsoas
- iliacus
- pectineus
- sartorius
- erector spinae

ANNOTATION KEY
Bold text = stretching muscles
* indicates deep muscle

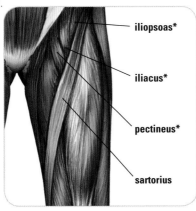

iliopsoas*

iliacus*

pectineus*

sartorius

Modification
Advanced: Continue from step 2, and straighten your leg, resting your flexed feet on your heels.

DOUBLE-LEG HINGE

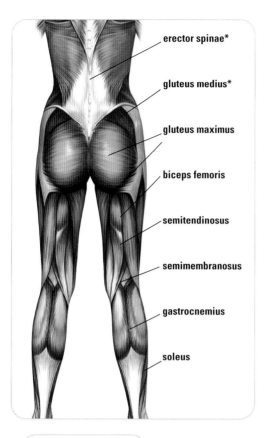

erector spinae*

gluteus medius*

gluteus maximus

biceps femoris

semitendinosus

semimembranosus

gastrocnemius

soleus

ANNOTATION KEY
Bold text = stretching muscles
* indicates deep muscle

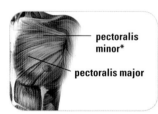

pectoralis minor*

pectoralis major

1 Stand behind a chair with your legs and feet parallel and generously outside of shoulder width. Bend your knees very slightly, and tuck your pelvis slightly forward, lift your chest, and press your shoulders downward and back.

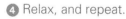

2 Bend forward from the hips, keeping your back flat, to grab the back of the chair. Your arms should form a 90-degree angle with your legs.

3 Try to bring your upper body closer to the floor while maintaining the position of your hands.

4 Relax, and repeat.

BEST FOR

- pectoralis major
- pectoralis minor
- gluteus maximus
- gluteus medius
- biceps femoris
- semitendinosus
- semimembranosus
- erector spinae
- gastrocnemius
- soleus

TARGETS
- Chest
- Hamstrings
- Lower back
- Gluteal area
- Calves

DO
- Keep your chest elevated.
- Exhale as you bend forward from the hips.

AVOID
- Tensing your neck or shoulders.
- Holding unnecessary tension in the upper body—relax and breathe in and out naturally.

SUPPORTED HAMSTRINGS STRETCH

1 Stand in front of a chair with your legs and feet parallel and shoulder-width apart. Bend your knees very slightly, and tuck your pelvis slightly forward, lift your chest, and press your shoulders downward and back.

BEST FOR

- biceps femoris
- semitendinosus
- semimembranosus
- erector spinae
- gastrocnemius
- soleus

TARGETS
- Hamstrings
- Calves

DO
- Keep your back straight and your chest open and lifted.

AVOID
- Bouncing to reach a deeper stretch.

2 Place your right leg on the seat of the chair.

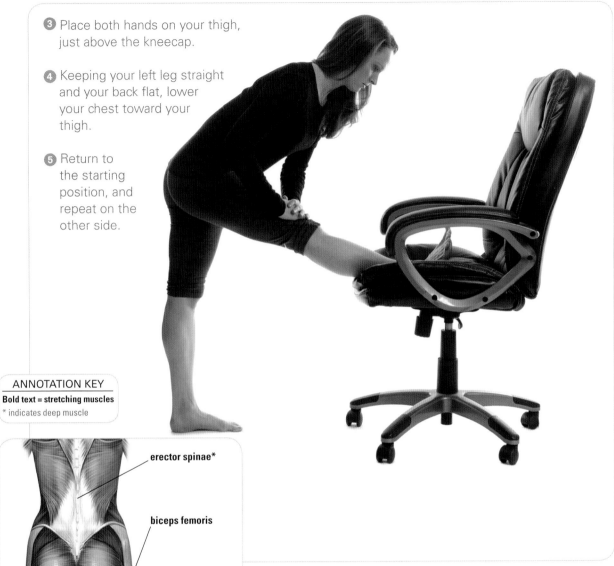

③ Place both hands on your thigh, just above the kneecap.

④ Keeping your left leg straight and your back flat, lower your chest toward your thigh.

⑤ Return to the starting position, and repeat on the other side.

ANNOTATION KEY

Bold text = stretching muscles

* indicates deep muscle

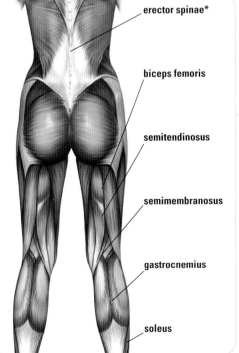

erector spinae*

biceps femoris

semitendinosus

semimembranosus

gastrocnemius

soleus

Modification

Advanced: From step 4, slide your hands down to your raised foot, aiming your chest toward your knee.

FOAM ROLLER STRETCHES

Here's a fact that may save you a great deal of time: When stretching a muscle with knots, you are only stretching the healthy muscle tissue. The knot remains intact. To relieve knotted muscles, massage is essential.

Of course nothing beats an expert massage from a professional massage therapist, but you can practice effective self-massage using a foam roller. This technique of using a foam roller on such trigger points is known as "self-myofascial release" (SMFR).

The Benefits of a Foam Roller

SMFR on a foam roller has many benefits, including:

- A foam roller allows you to control the amount of pressure you place on any trouble spots.
- A foam roller is relatively inexpensive.
- A small foam roller can travel with you.
- You can book a massage appointment with your foam roller anytime you wish!

Using a Foam Roller

To use it properly, you will need to control your body weight on the foam roller to generate the pressure necessary to break up the problematic spots, also known as "trigger points."

WHY MASSAGE?

Massaging a muscle (or group of muscles) prior to stretching has many benefits, including:

- The removal of metabolic waste, such as lactic acid, from the muscle. This will also aid in relieving post-exercise soreness and stiffness.
- Increases blood flow and improves circulation.
- Helps warm up your muscles.
- Relaxes your muscles.

- Roll back and forth across any stiff, painful areas for approximately 60 seconds, rest for 10 seconds, and repeat.
- Maintain proper stability (slight contraction) in your abdominals to protect your core (lower back, pelvis, and hips) during rolling.
- Breathe slowly and naturally throughout the entire exercise—this will help reduce any excess tension that the discomfort of rolling may cause.
- Avoid rolling over your bony areas.
- Perform the foam roller stretches included here three times a week to prevent stiffness and injury, and feel free to roll over any stiff or knotted areas two to three times a day. You can also use the roller on your trigger points before they knot up.
- Follow up with proper stretches that target the muscles you focused on during your foam roller exercises and massages.

MYOFASCIAL BALL THERAPY

Myofascial ball therapy may be the most inexpensive physical therapy instrument you can invest in. A small ball, such as a tennis ball, travels well and can provide a discreet massage, even in a work environment.

There are many reasons to use a ball for massage: it can relieve pain and physical stress, aid the body in becoming more limber, and help relieve foot and calf muscle cramps (especially for women who wear uncomfortably high heels).

Consider utilizing a tennis ball to massage your calves, hamstrings, gluteals, quadriceps, and back.

TENNIS BALL FOOT MASSAGE

❶ Sit comfortably on a chair. Place a ball on the bottom center area of your right foot, under the arch.

❷ Roll the ball back and forth from the ball of the foot all the way to the heel, and then roll it back and forth under the arch of your foot. Give careful attention to any areas that are uncomfortable. Continue for 60 seconds, rest, and then repeat.

❸ Switch feet, and repeat the sequence.

ITB ROLL

① Kneel upright facing the foam roller, with your knees a few inches behind the roller.

EXPERT'S TIP

If you feel any pain in the targeted area, stop rolling and rest on the area for 30 to 45 seconds.

② Lean forward onto all fours, placing your hands about a foot in front of the roller.

BEST FOR

- rectus femoris
- vastus lateralis
- vastus medialis
- vastus intermedius
- tractus iliotibialis

TARGETS
- Iliotibial band
- Quadriceps

DO
- Adjust the weight you place on your hands and the foot of your bent leg to relieve some of the pressure on the muscle being rolled and to find the appropriate level of intensity.
- Place your elbows down on the floor for added support if necessary.
- Emphasize the outside of your thigh.

AVOID
- Holding your breath.

③ Reach forward to rest your upper thigh against the roller, bend your left knee up toward your right knee, and place as much of your foot on the floor as possible.

④ Tilt your body slightly to the right, adjusting your body weight to achieve the desired pressure on the upper thigh, rolling slowly down to just above your knee.

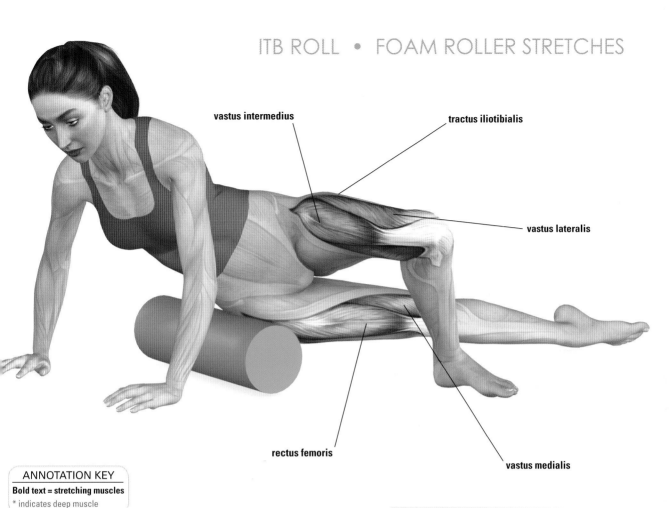

vastus intermedius

tractus iliotibialis

vastus lateralis

rectus femoris

vastus medialis

ANNOTATION KEY
Bold text = stretching muscles
* indicates deep muscle

⑤ Pause over uncomfortable areas before rolling back and forth over them until you feel some relief. Continue rolling for 60 seconds. Rest, and then repeat.

⑥ Switch sides, and repeat the entire sequence.

WHAT IS THE ITB?

The iliotibial band, usually just called the "ITB," is a thick band of fibrous tissue that runs down the outside of the leg, beginning at the hip and extending to the outer side of the tibia, just below the knee joint. The ITB functions in coordination with several of the thigh muscles to provide stability to the outside of the knee joint.

Many dancers, runners, cyclists, hikers, and other athletes experience a common thigh injury called iliotibial band syndrome. The repeated flexion and extension of the knee that all of these activities demand may cause the ITB area to become inflamed, producing hip and knee pain. Foam roller massage may be helpful in preventing and relieving the discomfort produced by this syndrome.

FOAM ROLLER LAT STRETCH

1 Kneel with your buttocks resting on your heels, and then shift your weight to the right so that your bent right leg is in front of the left. Place your hands on the foam roller in front of you.

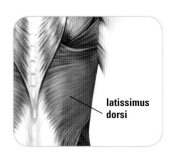

latissimus
dorsi

ANNOTATION KEY

Bold text = stretching muscles
* indicates deep muscle

2 Extend your right arm over the roller, tilt your body to the right, and gently lean your upper body down onto the foam roller.

TARGETS
• Latissimus dorsi

DO
• Keep control of your core, and maintain tight gluteal muscles.

AVOID
• Holding your breath.

3 Flatten your right palm on the floor for support, and with your left hand on the foam roller, carefully lift up your lower body enough to allow the foam roller to roll over the broad muscle of your middle back.

4 Pause over uncomfortable areas before rolling back and forth over them until you feel some relief. Continue rolling for 60 seconds. Rest, and then repeat.

BEST FOR

• latissimus dorsi

FOAM ROLLER BACK STRETCH

1 Sit on the floor and extend your legs in front of you in parallel position. Bend your knees slightly, keeping your heels on the floor. Your feet should be shoulder-width apart, with both feet on the floor.

2 Place the foam roller behind your gluteals and lower-back region.

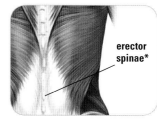

erector spinae*

ANNOTATION KEY

Bold text = stretching muscles

* indicates deep muscle

BEST FOR

- erector spinae

3 Lean back carefully onto the foam roller. Raise your hips slightly off the floor, lifting your buttocks upward as you simultaneously take small steps forward, allowing you to begin rolling the foam roller upward on your back.

TARGETS
- Back

DO
- Place your hands behind your head, extend them downward on the sides of your body, or wrap your arms around your chest.

AVOID
- Holding your breath.

4 Pause over uncomfortable areas before rolling back and forth over them until you feel some relief. Continue rolling for 60 seconds. Rest, and then repeat.

EXPERT'S TIP

To come out of this stretch at any time, lower your buttocks to the floor while taking baby steps forward.

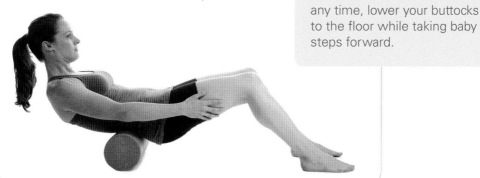

CALF AND HAMSTRINGS STRETCH

① Kneel upright with the foam roller in your hands, and then place the foam roller behind your knees.

② Carefully rock your pelvis slightly forward, just enough to place the foam roller deep behind your kneecaps.

③ Lower your body weight by sitting gently on the foam roller.

④ As you begin to sit, you will find that the foam roller will naturally move over your calf muscle. Guide the roller with your hands moving the roller slowly down toward your heels.

TARGETS
• Calves
• Hamstrings

DO
• Engage your core to adjust the amount of body weight you rest on the roller to find the appropriate level of intensity.

AVOID
• Leaning forward; keep your upper body upright.

EXPERT'S TIP

For a deeper, more advanced massage on the outer region of your calf muscles, place your hands on the very ends of the foam roller, and bend them slightly downward.

ANNOTATION KEY
Bold text = stretching muscles
* indicates deep muscle

5 Pause over uncomfortable areas before rolling back and forth over them until you feel some relief. Continue rolling for 60 seconds. Rest, and then repeat.

biceps femoris

semitendinosus

semimembranosus

gastrocnemius

soleus

BEST FOR

- gastrocnemius
- soleus
- biceps femoris
- semitendinosus
- semimembranosus

FOAM ROLLER SHIN STRETCH

1 Stand over the foam roller in a slight lunge, with your right leg in front and your leg left behind the foam roller. Place your hands on your right thigh just above the kneecap for support.

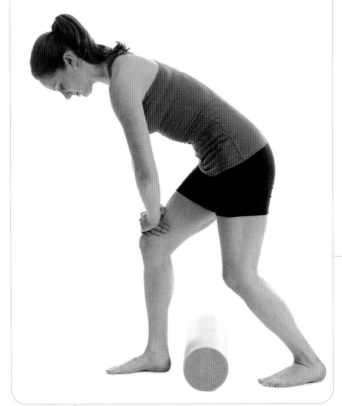

EXPERT'S TIP

For an advanced massage, rock your shin slightly back and forth, directly over the foam roller.

2 Lower your body downward, placing the upper region of your left shin just below the kneecap onto the foam roller. Place your hands on the floor in front of you.

TARGETS
• Shin

DO
• Control the amount of pressure you place on the foam roller by adjusting the amount of weight you place on your hands.

AVOID
• Holding your breath.

3 Sit down onto your calf muscle.

4 Pause over uncomfortable areas before rolling back and forth over them until you feel some relief. Continue rolling for 60 seconds. Rest, and then repeat.

5 Switch sides, and repeat the entire sequence.

BEST FOR

- tibialis anterior
- peroneus
- extensor digitorum

ANNOTATION KEY

Bold text = stretching muscles
* indicates deep muscle

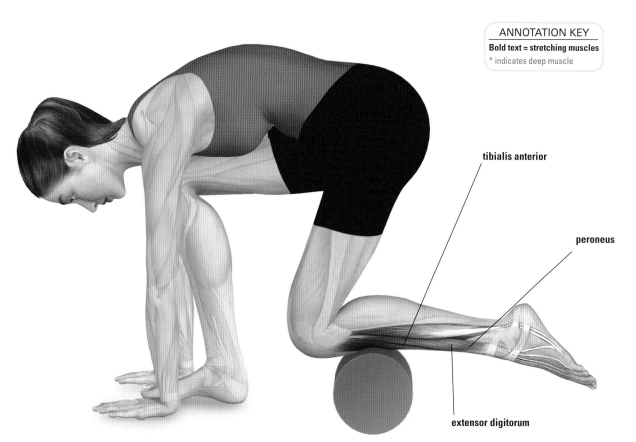

tibialis anterior

peroneus

extensor digitorum

EXTREME CHALLENGE

Not everyone has either the desire or the need to obtain this level of stretching ability—nor should they.

Still, there are many hobbies and professions that recommend or demand this form of control over the body, such as dancing or martial arts.

There is a myth that people with extreme flexibility, such as contortionists, age quickly: such "overuse" of the body will eventually render them immobile and inactive. The truth is that people who have achieved this form of flexibility typically live healthy, long, and very physically capable and active lives. In fact, extreme flexibility may decrease one's chances of suffering from osteoarthritis. Flexibility gives you strength and an improved range of motion, both positive benefits as you grow older.

Learn from Your Pet
An old adage states that the secret to a long and healthy life is a flexible spine. If you own a pet, observe how many times a day it stretches its back and limbs. Nearly all vertebrate animals exercise many times a day by stretching and flexing their bodies.

A Challenge for the Fit
Before you even attempt any of the Extreme Challenge stretches, perform the Stretching Session daily for a minimum of one month. And always consult your doctor before starting this or any stretching program. Keep in mind: you do not have to achieve this level of stretching ability to achieve success with this program—the Extreme Challenge is only for those who feel they have graduated to this level and have the personal desire or professional need to attain maximum flexibility.

FLEXIBLE OPTIONS

Some examples of sports and careers in which extreme flexibility is a benefit:

- figure skating
- hockey goaltending
- gymnastics
- diving
- dancing
- cheerleading
- track and field
- stunt performing
- mountain climbing
- martial arts

DANCER'S LUNGE

1 Kneel on all fours, with your knees directly below your hips. Position your hands on the floor just beyond shoulder-width apart, slightly in front of your body, palms downward and fingertips facing forward.

2 Keeping your left knee bent and your shin flat on the floor, lunge your right foot forward so that it stops just behind your right hand.

BEST FOR

- rectus femoris
- vastus lateralis
- vastus intermedius
- vastus medialis
- biceps femoris
- semitendinosus
- semimembranosus
- gluteus maximus
- gluteus medius
- gluteus minimus
- adductor longus
- adductor magnus
- adductor brevis
- gracilis
- pectineus
- obturator externus
- iliopsoas
- tensor fasciae latae

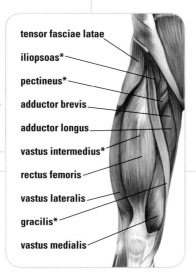

tensor fasciae latae
iliopsoas*
pectineus*
adductor brevis
adductor longus
vastus intermedius*
rectus femoris
vastus lateralis
gracilis*
vastus medialis

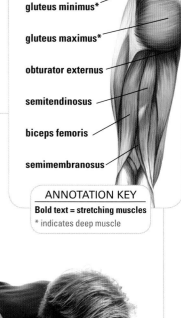

gluteus medius*
gluteus minimus*
gluteus maximus*
obturator externus
semitendinosus
biceps femoris
semimembranosus

ANNOTATION KEY
Bold text = stretching muscles
* indicates deep muscle

TARGETS
- Quadriceps
- Gluteal area
- Inner thighs
- Hamstrings
- Ball of the foot

DO
- Keep your back leg extended in line with your hips to form one long straight line.

AVOID
- Allowing your chest to sink.

3 Slide your right hand back and around your right ankle as you lower your chest to the floor so that your right thigh rests on your upper arm.

4 Bring your left arm out to the side, in line with the right, and push upward to straighten your left leg.

5 Release the stretch, switch legs, and repeat.

LYING-DOWN SIDE HAMSTRING STRETCH

1 Sit on the floor with your legs extended in front of you, bend your right knee, and wrap a small towel or other strap around the bottom of your right foot.

2 Grasp both ends of the towel with your left hand, and place your right hand behind you.

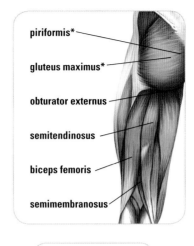

piriformis*
gluteus maximus*
obturator externus
semitendinosus
biceps femoris
semimembranosus

ANNOTATION KEY
Bold text = stretching muscles
* indicates deep muscle

BEST FOR

- biceps femoris
- semitendinosus
- semimembranosus
- gluteus maximus
- piriformis
- obturator externus

3 Carefully lean back, guiding your right leg backward until you are lying supine on the floor with your right knee bent to your chest.

TARGETS
- Hamstrings
- Gluteal region

DO
- Keep the leg on the floor straight and long.

4 Extend your right arm away from your torso, palm facing down.

5 Using the towel to guide you, slowly straighten your right leg until it is fully extended, your knee crossing your upper arm.

AVOID
- Bouncing to extend your leg farther back than you can go without discomfort—only stretch as far as the limits of your own body.

6 Release the stretch, switch legs, and repeat.

BILATERAL QUAD STRETCH

1 Kneel with your buttocks resting lightly on your heels.

2 Place your hands flat on the floor behind you, with your fingers pointing forward. Keep a slight bend in your elbows.

3 Lean slightly back to increase the intensity of the stretch.

BEST FOR

- transversus abdominis
- rectus abdominis
- rectus femoris
- vastus intermedius
- vastus lateralis
- sartorius
- pectineus
- iliopsoas
- tensor fasciae latae
- pectoralis major

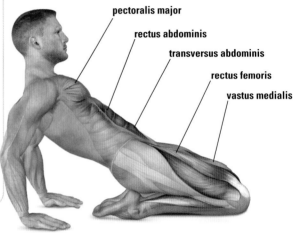

pectoralis major
rectus abdominis
transversus abdominis
rectus femoris
vastus medialis

TARGETS
- Abdominals
- Inner thighs
- Outer thighs
- Shins
- Chest

DO
- Contract and engage your gluteal muscles to avoid a curve in your lumbar spine; you will have space between the heels and the glutes.

AVOID
- Arching your back.

tensor fasciae latae
iliopsoas*
pectineus*
sartorius
vastus intermedius*

4 Continue to carefully lean backward until you are lying supine on the floor. Extend both arms away from your torso, palms facing up.

ANNOTATION KEY
Bold text = stretching muscles
* indicates deep muscle

FRONT SPLIT

1 Kneel on your left leg, with your right leg forward, making sure that your knee doesn't extend past your toes. Square your hips with your left knee flat on the floor.

2 Square your shoulders with your hands on the floor for balance.

tensor fasciae latae
iliopsoas*
pectineus*
adductor brevis
adductor longus

3 Carefully slide your right leg forward as you extend your left leg backward as far as you can without forcing the movement.

BEST FOR

- biceps femoris
- semitendinosus
- semimembranosus
- gluteus maximus
- gluteus medius
- gluteus minimus
- adductor longus
- adductor magnus
- adductor brevis
- gracilis
- pectineus
- obturator externus
- iliopsoas
- tensor fasciae latae

4 Sit up tall, and hold.

5 Release the stretch, switch legs, and repeat.

gluteus medius*
gluteus minimus*
gluteus maximus*
obturator externus
semitendinosus
biceps femoris
semimembranosus

TARGETS
- Inner thighs
- Hamstrings
- Gluteal region

DO
- Keep your chest open and lifted.

AVOID
- Forcing yourself downward too fast; perfect flat splits take practice.

ANNOTATION KEY
Bold text = stretching muscles
* indicates deep muscle

RUSSIAN SPLITS

RUSSIAN SPLIT

1 Sit upright, bringing the soles of your feet together.

2 Place your hands on the floor behind you.

BEST FOR

- biceps femoris
- semitendinosus
- semimembranosus
- gluteus maximus
- gluteus medius
- gluteus minimus
- adductor longus
- adductor magnus
- adductor brevis
- gracilis
- pectineus
- obturator externus
- iliopsoas
- tensor fasciae latae

3 Extend both legs out from the hips as far you can reach. Your feet should be slightly flexed so that your toes point upward.

TARGETS
- Hip adductors
- Hamstrings
- Inner thighs
- Gluteal region

DO
- Sit up as tall as possible, with your torso long.
- During the roll-through phase, stretch as far as you can, maintaining a flat back.

AVOID
- Bouncing your legs more widely open—you want to feel the stretch but never pain.

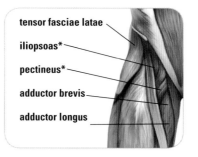

tensor fasciae latae
iliopsoas*
pectineus*
adductor brevis
adductor longus

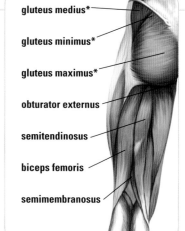

gluteus medius*
gluteus minimus*
gluteus maximus*
obturator externus
semitendinosus
biceps femoris
semimembranosus

ANNOTATION KEY
Bold text = stretching muscles
* indicates deep muscle

ROLL-THROUGH FROM RUSSIAN SPLIT

1 From the Russian Split, bring your hands forward, placing your palms on the floor in front of you.

2 Bring your chest toward the floor.

3 Extend your arms out to your sides as you bring your forehead to the floor.

4 Bring your legs behind you, and bend your arms to relax them above your head.

STANDING EXTENSIONS

ASSISTED SIDE TILT

1 Stand with your legs together, your feet turned out with heels touching. Place your right hand on the back of a chair or other stable object.

2 Shift your weight onto your right foot and bend your left leg upward, grasping the bottom of your left foot with your left hand.

TARGETS
- Hip adductors
- Hamstrings
- Inner thighs
- Gluteal region

DO
- Keep your back straight and your chest open.
- Pull your leg upward only as high as you can without hurting your hips.

AVOID
- Holding your breath.

3 Extend your left leg upward from the hip.

4 Keeping a firm grip on the bottom of your foot, tilt your upper body to the right until your torso and right leg form a 90-degree angle.

5 Release the stretch, switch legs, and repeat.

STANDING LEG EXTENSION

1 Stand with your legs and feet parallel and shoulder-width apart. Bend your right leg, grasping your ankle with your right hand and your toes with your left.

2 Keeping hold of your ankle and toes, hinge your right leg upward from the hip.

3 Continue stretching your leg upward until your right leg is fully extended perpendicular to the floor.

4 Release the stretch, switch legs, and repeat.

BEST FOR

- rectus femoris
- vastus lateralis
- vastus intermedius
- vastus medialis
- biceps femoris
- semitendinosus
- semimembranosus
- gluteus maximus
- gluteus medius
- gluteus minimus
- adductor longus
- adductor magnus
- adductor brevis
- gracilis
- pectineus
- obturator externus
- iliopsoas
- tensor fasciae latae

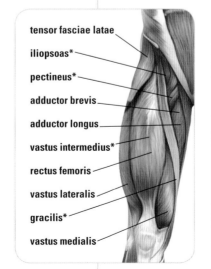

tensor fasciae latae
iliopsoas*
pectineus*
adductor brevis
adductor longus
vastus intermedius*
rectus femoris
vastus lateralis
gracilis*
vastus medialis

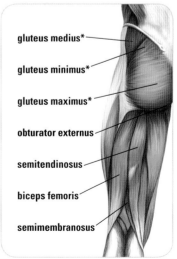

gluteus medius*
gluteus minimus*
gluteus maximus*
obturator externus
semitendinosus
biceps femoris
semimembranosus

ANNOTATION KEY
Bold text = stretching muscles
* indicates deep muscle

157

THE QUICK STRETCH PROGRAM

Taking a mere 10 to 15 minutes to complete, the Quick Stretch Program makes "not enough time" an invalid excuse for missing a daily stretch.

You know those 15 minutes in the morning you spend stumbling around sipping your coffee, just trying to get yourself moving? This 20-step program flows quickly and easily, taking advantage of bilateral stretches wherever possible to condense execution time. So start your day off right, and use those minutes to perform this program when you first wake up.

You can also take advantage of this program just before you go to bed. After a long day, your muscles often tighten up. Head to the bedroom 10 to 15 minutes early, and perform the Quick Stretch Program to relax your body and your mind, preparing both for a restful night's sleep.

Another great time to stretch is while watching TV—something most of us do every day. Instead of sinking into your sofa, sink into this program as you take in your favorite shows.

Even if it's impossible to find a block of 10 to 15 minutes, the Quick Stretch Program still works—you can break up the time throughout your day. Just follow the steps in order, holding for the indicated time or performing the indicated reps!

A FAMILY AFFAIR

Include your kids in this program, and make fitness a family affair! This quick program is short enough to keep the attention and interest of children of all ages.

1 Hold for **20 seconds**

Bilateral Seated Forward Bend
(pages 28–29)

2 Hold for **20 seconds**

Folded Butterfly Stretch
(page 31)

3 Hold for **20 seconds** each side

Lying-Down Pretzel Stretch
(pages 36–37)

4 Hold for **20 seconds** each side

Unilateral Knee-to-Chest Stretch
(page 38)

5 Hold for **20 seconds** each side

Lying-Down Figure 4 Stretch
(pages 42–43)

6 Hold for **20 seconds** each side

Side-Lying Rib Stretch
(pages 48–49)

7 Hold for **20 seconds**

Cobra Stretch
(pages 52–53)

8 Hold for **20 seconds** each side

Pigeon Stretch
(pages 56–57)

9 Hold for **20 seconds**

Shin Stretch
(pages 58–59)

10 Hold for **20 seconds**

Frog Straddle
(pages 60–61)

11 Hold for **20 seconds**

Toe Touch
(page 70)

12 Hold for **5 seconds** for 2 sets

Lion Stretch
(page 75)

13 Hold for **10 seconds** each side

Side Neck Tilt
(page 76)

14 Hold for **20 seconds**

Back-of-the-Neck Stretch
(page 77)

15 Hold for **20 seconds** each side

Triceps Stretch
(page 78)

16 Hold for **20 seconds** each side

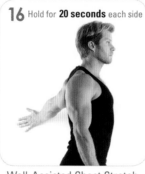

Wall-Assisted Chest Stretch
(pages 80–81)

17 Hold for **20 seconds** each side

Toe-Up Calf Stretch
(page 85)

18 Hold for **20 seconds** each side

Side-Lunge Stretch
(pages 92–93)

19 Hold for **20 seconds** each side

Forward Lunge
(pages 94–95)

20 Hold for **20 seconds**

Downward-Facing Dog
(pages 100–101)

CREDITS AND ACKNOWLEDGMENTS

All photographs by Jonathan Conklin/Jonathan Conklin Photography, Inc., except inset on page 148 by Aspen Photo/Shuttterstock.com.

All anatomical illustrations by Hector Aiza/3D Labz Animation India, except the insets on pages 74, 106, 107, 111, 115, 117, 119, 123, 124, 125, 126, 127, 132, 133, 134, 135, 137, 142, and 143 by Linda Bucklin/Shuttterstock.com.

Poster illustration by Linda Bucklin/Shutterstock; insets by Hector Aiza/3D Labz Animation India

Stylist: Brandon Liberati
Models: Craig Ramsay and Kelly Jacobs

ACKNOWLEDGMENTS

Thank you to my support team of family, friends, and colleagues: Brandon Liberati, Jerry Mitchell, Kelly Jacobs, Adam Jacobs, Catherine Wreford Ledlow, Kevin Rhodes, Amy Rivard, Gregg Simmons, Scott Barton, Barb Frederick (Innovative Artists), Scott Schwimer, Chuck and Lenore Ramsay (Dad and Mom), Phyllis Ramsay (Grandma), Scott and Vickie Davey, the town of Harrow, Ontario, and my Body Rehearsal clients.

I would also like to thank editor/designer Lisa Purcell, Moseley Road president Sean Moore, and photographer Jonathan Conklin.

The author and publisher also offer thanks to those closely involved in the creation of this book: Moseley Road president Sean Moore; editorial director Lisa Purcell; general manager Karen Prince; art director Brian MacMullen; editor Erica Gordon-Mallin; and designers Terasa Bernard and Danielle Scaramuzzo.

ABOUT THE AUTHOR

Internationally recognized in the fitness industry for his extensive knowledge and diverse expertise, CRAIG RAMSAY has more than 12 years of documented success transforming the health, bodies, and lives of his many clients. Craig's own versatile physical abilities were honed through his work on Broadway and as a ballet dancer, hockey player, fitness model, TV and film actor, and trained contortionist. A native of Ontario, Canada, who relocated to Los Angeles in 2008, Craig quickly established himself as a fitness and dance expert, model, and TV host. Successfully juggling his many careers, Craig's résumé has made him one of the most sought-after fitness experts in the business. He is well known for establishing a rapport with every client—from Hollywood's top celebrities to professional athletes—and proudly accepts his nickname of "World's Nicest Trainer."

Model KELLY JACOBS, a native of Wisconsin, began her dance training at age five. She received a psychology degree from the University of Wisconsin–Madison, but her passion for musical theater led her to a performing career that took her around the world, from Busch Gardens, Tampa, to Tokyo Disney to Holland America Cruise Lines. She made her Broadway debut serving as dance captain in Disney's *Mary Poppins*. Kelly also danced in *The Radio City Christmas Spectacular* at Radio City Music Hall for many years and appeared in *Mame* at the Kennedy Center in Washington, D.C. Other regional credits include *The Producers, Tommy, Camelot,* and *Cats.*